T0116225

CAMBRIDGE LIBRARY COLLECTION

Books of enduring scholarly value

Earth Sciences

In the nineteenth century, geology emerged as a distinct academic discipline. It pointed the way towards the theory of evolution, as scientists including Gideon Mantell, Adam Sedgwick, Charles Lyell and Roderick Murchison began to use the evidence of minerals, rock formations and fossils to demonstrate that the earth was older by millions of years than the conventional, Bible-based wisdom had supposed. They argued convincingly that the climate, flora and fauna of the distant past could be deduced from geological evidence. Volcanic activity, the formation of mountains, and the action of glaciers and rivers, tides and ocean currents also became better understood. This series includes landmark publications by pioneers of the modern earth sciences, who advanced the scientific understanding of our planet and the processes by which it is constantly re-shaped.

A Companion to the Mountain Barometer

The Englefield mountain barometer was designed to calculate altitude and was so easy to use that a reading could be taken out of the window of a carriage – provided the horses stood still. Using a bar of mercury, which fell in the lower air pressure of higher altitudes, the barometer gave readings accurate to one thousandth of an inch. By taking a mercury reading at two locations, the owner could work out the difference in altitude between them. In this companion book, first published in 1817, the maker of the new barometer, Thomas Jones, provides tables listing the heights of objects measuring between fifteen and thirty-one inches of mercury. These measurements account for the heights of all mountains in England. He also includes tables that show how to allow for the expansion of both air and mercury. A fascinating book for historical researchers and experimenters in physics alike.

Cambridge University Press has long been a pioneer in the reissuing of out-of-print titles from its own backlist, producing digital reprints of books that are still sought after by scholars and students but could not be reprinted economically using traditional technology. The Cambridge Library Collection extends this activity to a wider range of books which are still of importance to researchers and professionals, either for the source material they contain, or as landmarks in the history of their academic discipline.

Drawing from the world-renowned collections in the Cambridge University Library and other partner libraries, and guided by the advice of experts in each subject area, Cambridge University Press is using state-of-the-art scanning machines in its own Printing House to capture the content of each book selected for inclusion. The files are processed to give a consistently clear, crisp image, and the books finished to the high quality standard for which the Press is recognised around the world. The latest print-on-demand technology ensures that the books will remain available indefinitely, and that orders for single or multiple copies can quickly be supplied.

The Cambridge Library Collection brings back to life books of enduring scholarly value (including out-of-copyright works originally issued by other publishers) across a wide range of disciplines in the humanities and social sciences and in science and technology.

3

RULE.

1st. For the expansion of the mercury by the attached Thermometer, multiply the number in the Table corresponding with the coldest Barometer by the difference of the attached Thermometer, and add it to the coldest Barometer.

2d. For the approximate height.—From the second Table take the numbers corresponding with the inches and parts of the Barometers at the two stations, and the difference of those numbers will be the approximate height in feet.

3d. For the air or detached Thermometer.—Take from the third Table the number corresponding with the approximate height, multiply it by the mean of the detached Thermometers, and the product added to the approximate height will give the real height.

Method of taking an Observation, and computing the Altitude.

About five minutes before you arrive at the place of observation, take out the Thermometer (which is used as detached), holding it by the upper end at nearly arm's length, and in the shade of your person, if the sun shines; it very soon takes the temperature of the air, and is not sensibly affected by the hand. The degree being observed and written down as at D 7.4, and the Thermometer returned; place the Barometer in a vertical position, move the index until its under surface is made a tangent to the convex surface of the mercury; read off and register the inches and parts as B 29.463, together with the attached Thermometer as at A 10; then proceed to the next station and repeat the operations, as D 5.2, B 25.678, and A 7.8. Now look to the first Table, and opposite 25 inches (the coldest Barometer) is .0050, which multiplied by 2.2 (the difference of the attached Thermometers) gives .011 to be added to 25.678 = 25.689. Next look in the left hand column of the second Table for 29.463, and the number 1325 will be found to correspond with it, which set down as at H: then find the number opposite 25.689, which is 4897, and place it under the former: now, by subtracting the one from the other we have the approximate height, 3572. Take from the third Table the number corresponding with the nearest number to the approximate height, which in this case is 15.3, and multiply it by the mean of the detached Thermometer 6.3, and you have 96.39 to add, which gives you 3668.39 feet, the real height.

EXAMPLE.

B	H	A	D
29.463	1325	10	7.4
25.678		7.8	5.2
.011			
25.689	4897	2.2	12.6
		.0050	
	3572	0100	6.3
	96.39	0100	15.3
	3668.39	.01100	459
			918
			96.39

TABLE I.

For the Expansion of the Mercury.

31	.0062
30	.0060
29	.0058
28	.0056
27	.0054
26	.0052
25	.0050
24	.0048
23	.0046
22	.0044
21	.0042
20	.0040
19	.0038
18	0036
17	.0034
16	.0032
15	.0030

A DESCRIPTION AND USE

OF

THE ENGLEFIELD MOUNTAIN BAROMETER.

THE Barometer tube is about 32¼ inches in length; its bore is from a tenth of an inch to two-tenths in diameter, and external diameter is three-tenths of an inch. This sized bore is fully sufficient to allow the free motion of the mercury. The cistern is of box-wood, turned truly cylindrical, and is one inch in its internal diameter, and an inch in depth; a short stem projects from its top (the instrument being in a position for making an observation), for the purpose of giving a firmer hold to the tube: this stem is perforated with a hole sufficiently large to admit the tube, which is glued to it in the usual mode. This tube projects into the cistern exactly to half its depth. The bottom of the cistern is closed by a brass ring and a leather bag, which screws on the cistern. The tube being filled in

A DESCRIPTION

OF

THE TABLES.

In addition to the various attempts made to simplify and render easy the operation of computing barometrical altitudes that have been laid before the public, I beg leave to present the following Tables, trusting they will not be found deficient in accuracy to any mode at present known.

Perhaps it may not be deemed intrusive to observe, that the principal inducement which led to the composition has been the great encouragement I have been honoured with in the manufacture and sale of Mountain Barometers, particularly the Englefield, having sold between three and four hundred since their first introduction, independent of those of the former construction.

In constructing these Tables, I have endeavoured that their results should correspond as nearly as possible with those obtained from the generally-approved formulæ of Dr. Maskelyne.

The first Table in use is that for the expansion of the mercury in the Barometer as given by the attached Thermometer, and which is constructed agreeably to the rule given by Sir George Shuckburgh.

The second Table consists of 32 pages computed from the logarithms to every thousandth of an inch, from 15 to 31 inches, and gives the approximate height in feet by the difference of the two numbers corresponding with the observed heights at the two stations.

The third Table consists of two pages constructed on the well-known principles of the expansion of air, and is for the correction of the air or detached Thermometers.

A

COMPANION

TO THE

MOUNTAIN BAROMETER,

CONSISTING OF TABLES,

WHEREBY THE OPERATION OF COMPUTING HEIGHTS WITH
THAT INSTRUMENT IS RENDERED EXTREMELY SIMPLE
AND EASY, WHILST ITS ACCURACY IS INFERIOR TO NO
OTHER MODE:

TOGETHER WITH

A DESCRIPTION AND USE

OF THE

ENGLEFIELD MOUNTAIN BAROMETER.

———

By THOMAS JONES,

ASTRONOMICAL AND PHILOSOPHICAL INSTRUMENT MAKER
TO HIS ROYAL HIGHNESS THE DUKE OF CLARENCE.

———

LONDON:

Printed by Richard and Arthur Taylor, Shoe Lane,

FOR THE AUTHOR, AND SOLD BY HIM AT HIS MANUFACTORY,
NO. 62, CHARING-CROSS.

1817.

Price 2s. 6d.

CAMBRIDGE UNIVERSITY PRESS

Cambridge, New York, Melbourne, Madrid, Cape Town,
Singapore, São Paolo, Delhi, Mexico City

Published in the United States of America by Cambridge University Press, New York

www.cambridge.org
Information on this title: www.cambridge.org/9781108049375

© in this compilation Cambridge University Press 2012

This edition first published 1817
This digitally printed version 2012

ISBN 978-1-108-04937-5 Paperback

A Companion
to the
Mountain Barometer

THOMAS JONES

CAMBRIDGE
UNIVERSITY PRESS

the usual way, and the instrument held inverted in a perpendicular position, mercury is poured into the cistern till
it is filled within two-tenths of an inch of the top. The bag
is then firmly screwed on. The essential part of the instrument is now finished. The end of the tube in the cistern
can never be uncovered by the mercury in any possible position, and of course no air can ever enter into it; and, as
the areas of the cistern and tube are as the squares of the
diameters, the diameter of the bore of the tube being 1,
its external diameter 3, and the diameter of the cistern
1.0, the area of the cistern is $100-9=91$; and there being
two-tenths of an inch left empty in the cistern, the mercury must fall 182-tenths, or 18 inches and two-tenths, before the cistern is quite full; a space adequate to the measure of greater heights than any known mountain on the
earth, much more so to any height in this country. It will
not easily be believed, by those who have not seen it, that
the air will act on a cistern thus completely closed, and of
which the wood, in its thinnest part, is above a quarter of
an inch in thickness; but the fact is, that when the instrument is suspended by the side of the Mountain Barometer of the best construction, with an open cistern,
no difference whatever can be perceived in their sensibility
to the variations of the atmosphere. It is obvious that the
variations of altitude, in this instrument of dimensions,
above stated, will be one ninety-first part less than in a
Barometer furnished with an apparatus for bringing the
surface of the mercury in the cistern to a fixed level: this
defect is remedied by measuring the content of every tube
separately, and marking the correction with a diamond near
the top of the glass tube in front of the instrument. They
are marked $\frac{1}{44}, \frac{1}{53}, \frac{1}{66}, \frac{1}{70}$, &c. according to the bores of
the tubes, and the quantity is always to be added; that is,
one-44th, one-53d, one-66th, one-70th, &c. of the re-
sult found by the usual methods, is to be added to that re-
sult, by which method it is presumed that all errors from
the want of a gauge-point must be prevented. Indeed a
moment's consideration will convince us that the gauge-
point corrections of the best instruments of the old construction are very doubtful, on account of the great incon-
venience of setting them, even in their very best mode of
construction.

The tube and cistern being thus prepared, are mounted
in a mahogany tube of the size of a common walking-stick;
the stem of the cistern enters the mahogany tube, and the
cistern is completely covered and secured by a brass tube,

containing the bottom screw, which presses against the
bag for making the instrument portable.

For the observation of the height of the mercury, two
opposite slits are cut in the mahogany tube, reaching from
about 32 to 20 inches for the long scales ; and 32 to 25
inches for the short ones; which are sufficiently long for
any purpose in this country. The front slit has its sides
bevelled, and is exteriorly about three-fourths of an inch
wide ; on one side is fixed a brass plate, divided as usual
into inches, tenths, and twentieths. On this plate a nonius
slides, moveable by a small piece, which reads off, as in
other Barometers, to 1000dth of an inch. To this nonius a
small portion of brass tube is attached, which embraces the
Barometer tube, and its lower edge is, in observation, made
a tangent to the convex surface of the mercury, as in other
well-constructed Barometers ; and the very narrow slit
behind gives abundant light for observation.

On the bevelled side of the front slit, opposite the scale,
a Thermometer is placed for taking the heat of the instrument; which is so contrived as to take out of its place,
and answer the purpose of the attached and detached Thermometer.

A thin brass tube, with slits in it, turns half round, in
the usual manner, and covers the apertures above described
in the mahogany tube when the Barometer is not in use.

The mahogany tube is made rather tapering, and with a
ferrule at the end opposite the cistern. This ferrule unscrews,
and shows a steel ring, by which the Barometer may be
suspended when convenient: likewise a small milled head
(in the best Barometers) for giving motion to the nonius
by means of a screw.

The Method of taking an Observation and computing the Altitude.

Having thus described the instrument, a few practical
remarks on the manner of using it may not be superfluous.

When I am about to make an observation, about five
minutes before I arrive at the place I take out the Thermometer, holding it by the upper end at nearly arm's length
from my body, and, if the sun shines, in the shade of my
person. It very soon takes the temperature of the air, and
is not sensibly affected by the heat of the hand. The heat
being observed and written down as at D 7.4, and the Thermometer returned, the Barometer is turned up, the bottom
screw unscrewed as far as it will come, the brass tube half
turned, and the instrument held between the finger and

thumb of the left hand above the slit, so as to let it hang freely in a perpendicular position. Few persons, if any, have sufficient steadiness of hand to prevent little vibrations in the mercury in this position: the hand, therefore, should be either rested against any fixed body, or, if no such occurs, by kneeling on one knee. The cistern should be let down so as to touch the ground, the left hand holding the Barometer in a vertical position, which a little practice will render very easy. The index must then be moved by the knob or head at the ferrule, till its under surface, as before stated, is tangent to the mercury. A few light taps should be given to the tube, to ascertain that the mercury has fallen as low as it can. The height being then read off and registered as at B 29.463, together with that of the attached Thermometer as A 10, the brass tube is turned back, so as to cover the slits; the instrument gently inverted, and the whole is finished. All this may be done in two minutes.

It may be just mentioned that when the Barometer is carried by a careful person, it is by no means necessary to screw up the bag between every station. Now proceed to the next station, and repeat the operation and register as at D 5.2, B 25.727, and A 7.8. Now look to the first Table, and opposite 25 inches (the coldest Barometer) is .0050, which multiplied by 2.2 (the difference of the attached Thermometer) gives .011 to be added to 25.727 = 25.738. Next look in the left-hand column of the second Table for 29.463, and the number 1325 will be found to correspond with it, which set down as at H: then find the number opposite 25.738, which is 4847, and place it under the former. Now, by subtracting the one from the other, we have the approximate height 3522. Take from the third Table the number corresponding to the approximate height, which in this case is 15, and multiply it by the mean of the detached Thermometer 6.3, and you have 94.5 to add, making 3616.5, which divided by 70 (the number on the glass tube for the correction of the cistern in this case) gives 51.6 to be added, making 3668.1 the real height.

It may not be improper here to add, that I have found by experience that it is not necessary to quit the chaise in order to make observations with this Barometer; it is only requisite for the horses to stand still. The Thermometer, if held at arm's length out of the chaise window, will give the temperature exactly, before the order is given to stop the carriage, and the delay to the traveller will not much exceed a minute, as the observation may be read off and written down while the carriage is again going on.

When the Barometer is intended to be made portable by means of the bottom screw (at the cistern) which is covered with a brass cap to prevent its being disturbed, great care should be taken not to force the mercury to the top of the glass tube, as it would be liable to burst the bag and let the mercury run out, thereby injuring the instrument. When the bottom screw is screwed up, and the Barometer in a vertical position with the cistern downwards, the mercury should stand about one-eighth of an inch from the top of the glass tube. In carrying the Barometer, great care should be taken to keep the cistern above the horizontal position of the instrument, at about an angle of forty-five degrees; it may be carried under the arm, or in the hand at arm's length.

As it may be desirable sometimes to compare British with foreign Barometrical measurement, or *vice versâ*, an example or two may not be deemed unacceptable. Suppose it were required to convert the altitude 3668.1 feet into French measure: The metre being 3.281 English feet;

say, As 3.281 : 1 : : 3668.1 : 1117.982. M. de Humboldt having given the height of Guanaxuato above the level of the

sea 2084.3 metres; say, As 1 : 3.281 : : 2084.3 : 6838.588: and in like manner any other measure may be compared.

B	H	A	D
29.463	1325	10	7.4
25.727		7.8	5.2
.011			
	4847	2.2	12.6
25.738		.0050	
	3522		6.3
	94.5	.0100	15.
		.0100	
70	3616.5		45.
		.01100	90.
	51.6		
			94.5
	3668.1		

In.dec.	0	1	2	3	4	5	6	7	8	9
		In. dec. 15.000					In. dec. 15.499			
15.00	18916	18915	18913	18912	18910	18908	18906	18904	18902	18901
.01	18899	18898	18896	18894	18892	18890	18888	18886	18884	18883
.02	18881	18880	18878	18877	18875	18873	18871	18869	18867	18866
.03	18864	18863	18861	18860	18858	18856	18854	18852	18850	18849
.04	18847	18846	18844	18842	18840	18838	18836	18834	18832	18831
.05	18829	18828	18826	18825	18823	18821	18819	18817	18815	18814
.06	18812	18811	18809	18808	18806	18804	18802	18800	18798	18797
.07	18795	18794	18792	18791	18789	18787	18785	18783	18781	18780
.08	18778	18777	18775	18773	18771	18769	18767	18765	18763	18762
.09	18760	18759	18757	18756	18754	18752	18750	18748	18746	18745
.10	18743	18742	18740	18739	18737	18735	18733	18731	18729	18728
.11	18726	18725	18723	18722	18720	18718	18716	18714	18712	18711
.12	18709	18708	18706	18704	18702	18700	18698	18696	18694	18693
.13	18691	18690	18688	18687	18685	18683	18681	18679	18677	18676
.14	18674	18673	18671	18670	18668	18666	18664	18662	18660	18659
.15	18657	18656	18654	18653	18651	18649	18647	18645	18643	18642
.16	18640	18639	18637	18636	18634	18632	18630	18628	18626	18625
.17	18623	18622	18620	18618	18616	18614	18612	18610	18608	18607
.18	18605	18604	18602	18601	18599	18597	18595	18593	18591	18590
.19	18588	18587	18585	18584	18582	18580	18578	18576	18574	18573
.20	18571	18570	18568	18567	18565	18563	18561	18559	18557	18556
.21	18554	18553	18551	18550	18548	18546	18544	18542	18540	18539
.22	18537	18536	18534	18533	18531	18529	18527	18525	18523	18522
.23	18520	18519	18517	18516	18514	18512	18510	18508	18506	18505
.24	18503	18502	18500	18498	18496	18494	18492	18490	18488	18487
15.25	18485	18484	18482	18481	18479	18477	18475	18473	18471	18470
.26	18468	18467	18465	18464	18462	18460	18458	18456	18454	18453
.27	18451	18450	18448	18447	18445	18443	18441	18439	18437	18436
.28	18434	18433	18431	18430	18428	18426	18424	18422	18420	18419
.29	18417	18416	18414	18413	18411	18409	18407	18405	18403	18402
.30	18400	18399	18397	18396	18394	18392	18390	18388	18386	18385
.31	18383	18382	18380	18379	18377	18375	18373	18371	18369	18368
.32	18366	18365	18363	18362	18360	18358	18356	18354	18352	18351
.33	18349	18348	18346	18345	18343	18341	18339	18337	18335	18334
.34	18332	18331	18329	18328	18326	18324	18322	18320	18318	18317
.35	18315	18314	18312	18311	18309	18307	18305	18303	18301	18300
.36	18298	18297	18295	18294	18292	18290	18288	18286	18284	18283
.37	18281	18280	18278	18277	18275	18273	18271	18269	18267	18266
.38	18264	18263	18261	18260	18258	18256	18254	18252	18250	18249
.39	18247	18246	18244	18243	18241	18239	18237	18235	18233	18232
.40	18230	18229	18227	18226	18224	18222	18220	18218	18216	18215
.41	18213	18212	18210	18209	18207	18205	18203	18201	18199	18198
.42	18196	18195	18193	18192	18190	18188	18186	18184	18182	18181
.43	18179	18178	18176	18175	18173	18171	18169	18167	18165	18164
.44	18162	18161	18159	18158	18156	18154	18152	18151	18149	18148
.45	18146	18145	18143	18142	18140	18138	18136	18134	18132	18131
.46	18129	18128	18126	18125	18123	18121	18119	18117	18115	18114
.47	18112	18111	18109	18108	18106	18104	18102	18100	18098	18097
.48	18095	18094	18092	18091	18089	18087	18085	18083	18081	18080
15.49	18078	18077	18075	18074	18072	18070	18068	18067	18065	18064
In.dec.	0	1	2	3	4	5	6	7	8	9

In.dec.	In. dec. 15.500					In. dec. 15.999				
In.dec.	0	1	2	3	4	5	6	7	8	9
15.50	18062	18061	18059	18058	18056	18054	18052	18050	18048	18047
.51	18045	18044	18042	18041	18039	18037	18035	18033	18031	18030
.52	18028	18027	18025	18024	18022	18020	18018	18016	18014	18013
.53	18011	18010	18008	18007	18005	18003	18001	17999	17997	17996
.54	17994	17993	17991	17990	17988	17986	17984	17983	17981	17980
.55	17978	17977	17975	17974	17972	17970	17968	17966	17964	17963
.56	17961	17960	17958	17957	17955	17953	17951	17949	17947	17946
.57	17944	17943	17941	17940	17938	17936	17934	17932	17930	17929
.58	17927	17926	17924	17923	17921	17919	17917	17916	17914	17913
.59	17911	17910	17908	17907	17905	17903	17901	17899	17897	17896
.60	17894	17893	17891	17890	17888	17886	17884	17883	17881	17880
.61	17878	17877	17875	17874	17872	17870	17868	17866	17864	17863
.62	17861	17860	17858	17857	17855	17853	17851	17849	17847	17846
.63	17844	17843	17841	17840	17838	17836	17834	17832	17830	17829
.64	17827	17826	17824	17823	17821	17819	17817	17816	17814	17813
.65	17811	17810	17808	17807	17805	17803	17801	17799	17797	17796
.66	17794	17793	17791	17790	17788	17786	17784	17782	17780	17779
.67	17777	17776	17774	17773	17771	17769	17767	17766	17764	17763
.68	17761	17760	17758	17757	17755	17753	17751	17749	17747	17746
.69	17744	17743	17741	17740	17738	17736	17734	17733	17731	17730
.70	17728	17727	17725	17724	17722	17720	17718	17716	17714	17713
.71	17711	17710	17708	17707	17705	17703	17701	17700	17698	17697
.72	17695	17694	17692	17691	17689	17687	17685	17683	17681	17680
.73	17678	17677	17675	17674	17672	17670	17668	17667	17665	17664
.74	17662	17661	17659	17658	17656	17654	17652	17650	17648	17647
15.75	17645	17644	17642	17641	17639	17637	17635	17633	17631	17630
.76	17628	17627	17625	17624	17622	17620	17618	17617	17615	17614
.77	17612	17611	17609	17608	17606	17604	17602	17600	17598	17597
.78	17595	17591	17592	17591	17589	17587	17585	17584	17582	17581
.79	17579	17578	17576	17575	17573	17571	17569	17567	17565	17564
.80	17562	17561	17559	17558	17556	17554	17552	17551	17549	17548
.81	17546	17545	17543	17542	17540	17538	17536	17534	17532	17531
.82	17529	17528	17526	17525	17523	17521	17519	17518	17516	17515
.83	17513	17512	17510	17509	17507	17505	17503	17501	17499	17498
.84	17496	17495	17493	17492	17490	17488	17486	17485	17483	17482
.85	17480	17479	17477	17476	17474	17472	17470	17469	17467	17466
.86	17464	17463	17461	17460	17458	17456	17454	17452	17450	17449
.87	17447	17446	17444	17443	17441	17439	17437	17436	17434	17433
.88	17431	17430	17428	17427	17425	17423	17421	17419	17417	17416
.89	17414	17413	17411	17410	17408	17406	17404	17403	17401	17400
.90	17398	17397	17395	17394	17392	17390	17388	17386	17384	17383
.91	17381	17380	17378	17377	17375	17373	17371	17370	17368	17367
.92	17365	17364	17362	17361	17359	17357	17355	17353	17351	17350
.93	17348	17347	17345	17344	17342	17340	17338	17337	17335	17334
.94	17332	17331	17329	17328	17326	17324	17322	17321	17319	17318
.95	17316	17315	17313	17312	17310	17308	17306	17305	17303	17302
.96	17300	17299	17297	17296	17294	17292	17290	17289	17287	17286
.97	17284	17283	17281	17280	17278	17276	17274	17272	17270	17269
.98	17267	17266	17264	17263	17261	17259	17257	17256	17254	17253
15.99	17251	17250	17248	17247	17245	17243	17241	17239	17237	17236
In.dec.	0	1	2	3	4	5	6	7	8	9

In.dec.	0	1	2	3	4	5	6	7	8	9
		In. dec. 16.000					In. dec. 16.499			
16.00	17234	17233	17231	17230	17228	17226	17224	17223	17221	17220
.01	17218	17217	17215	17214	17212	17210	17208	17207	17205	17204
.02	17202	17201	17199	17198	17196	17194	17192	17191	17189	17188
.03	17186	17185	17183	17182	17180	17178	17176	17175	17173	17172
.04	17170	17169	17167	17166	17164	17162	17160	17158	17156	17155
.05	17153	17152	17150	17149	17147	17145	17143	17142	17140	17139
.06	17137	17136	17134	17133	17131	17129	17127	17125	17123	17122
.07	17120	17119	17117	17116	17114	17112	17110	17109	17107	17106
.08	17104	17103	17101	17100	17098	17096	17094	17093	17091	17090
.09	17088	17087	17085	17084	17082	17080	17078	17077	17075	17074
.10	17072	17071	17069	17068	17066	17064	17062	17061	17059	17058
.11	17056	17055	17053	17052	17050	17048	17046	17045	17043	17042
.12	17040	17039	17037	17036	17034	17032	17030	17029	17027	17026
13	17024	17023	17021	17020	17018	17016	17014	17013	17011	17010
.14	17008	17007	17005	17004	17002	17000	16998	16996	16994	16993
.15	16991	16990	16988	16987	16985	16983	16981	16980	16978	16977
.16	16975	16974	16972	16971	16969	16967	16965	16964	16962	16961
.17	16959	16958	16956	16955	16953	16951	16949	16948	16946	16945
.18	16943	16942	16940	16939	16937	16935	16933	16932	16930	16929
.19	16927	16926	16924	16923	16921	16919	16917	16916	16914	16913
.20	16911	16910	16908	16907	16905	16903	16901	16900	16898	16897
.21	16895	16894	16892	16891	16889	16887	16885	16884	16882	16881
.22	16879	16878	16876	16875	16873	16871	16869	16867	16865	16864
.23	16862	16861	16859	16858	16856	16854	16852	16851	16849	16848
.24	16846	16845	16843	16842	16840	16838	16836	16835	16833	16832
16.25	16830	16829	16827	16826	16824	16822	16820	16819	16817	16816
.26	16814	16813	16811	16810	16808	16806	16804	16803	16801	16800
.27	16798	16797	16795	16794	16792	16790	16788	16787	16785	16784
.28	16782	16781	16779	16778	16776	16774	16772	16771	16769	16768
.29	16766	16765	16763	16762	16760	16758	16756	16755	16753	16752
.30	16750	16749	16747	16746	16744	16742	16740	16739	16737	16736
.31	16734	16733	16731	16730	16728	16726	16724	16723	16721	16720
.32	16718	16717	16715	16714	16712	16710	16708	16707	16705	16704
.33	16702	16701	16699	16698	16696	16694	16692	16691	16689	16688
.34	16686	16685	16683	16682	16680	16678	16676	16675	16673	16672
.35	16670	16669	16667	16666	16664	16662	16660	16659	16657	16656
.36	16654	16653	16651	16650	16648	16646	16644	16643	16641	16640
.37	16638	16637	16635	16634	16632	16631	16629	16628	16626	16625
.38	16623	16621	16619	16618	16616	16614	16612	16611	16610	16609
.39	16607	16605	16603	16602	16600	16598	16596	16595	16593	16592
.40	16591	16590	16588	16587	16585	16583	16581	16580	16578	16577
.41	16575	16574	16572	16571	16569	16567	16565	16564	16562	16561
.42	16559	16558	16556	16555	16553	16551	16549	16548	16546	16545
.43	16543	16542	16540	16539	16537	16535	16533	16532	16530	16529
.44	16527	16526	16524	16523	16521	16519	16517	16516	16514	16513
.45	16511	16510	16508	16507	16505	16503	16501	16500	16498	16497
.46	16495	16494	16492	16491	16489	16488	16486	16485	16483	16482
.47	16480	16479	16477	16476	16474	16472	16470	16469	16467	16466
.48	16464	16463	16461	16460	16458	16456	16454	16453	16451	16450
16.49	16448	16447	16445	16444	16442	16440	16438	16437	16435	16434
In.dec.	0	1	2	3	4	5	6	7	8	9

In.dec.	In. dec. 16.500					In. dec. 16.999				
	0	1	2	3	4	5	6	7	8	9
16.50	16432	16431	16429	16428	16426	16424	16422	16421	16419	16418
.51	16416	16415	16413	16412	16410	16409	16407	16406	16404	16403
.52	16401	16400	16398	16397	16395	16393	16391	16390	16388	16387
.53	16385	16384	16382	16381	16379	16377	16375	16374	16372	16371
.54	16369	16368	16366	16365	16363	16361	16359	16358	16356	16355
.55	16353	16352	16350	16349	16347	16346	16344	16343	16341	16340
.56	16338	16337	16335	16334	16332	16330	16328	16327	16325	16324
.57	16322	16321	16319	16318	16316	16314	16312	16311	16309	16308
.58	16306	16305	16303	16302	16300	16298	16296	16295	16293	16292
.59	16290	16289	16287	16286	16284	16283	16281	16280	16278	16277
.60	16275	16274	16272	16271	16269	16267	16265	16264	16262	16261
.61	16259	16258	16256	16255	16253	16251	16249	16248	16246	16245
.62	16243	16242	16240	16239	16237	16236	16234	16233	16231	16230
.63	16228	16227	16225	16224	16222	16220	16218	16217	16215	16214
.64	16212	16211	16209	16208	16206	16205	16203	16202	16200	16199
.65	16197	16196	16194	16193	16191	16189	16187	16186	16184	16183
.66	16181	16180	16178	16177	16175	16173	16171	16170	16168	16167
.67	16165	16164	16162	16161	16159	16157	16155	16154	16152	16151
.68	16149	16148	16146	16145	16143	16142	16140	16139	16137	16136
.69	16134	16133	16131	16130	16128	16126	16124	16123	16121	16120
.70	16118	16117	16115	16114	16112	16110	16108	16107	16105	16104
.71	16102	16101	16099	16098	16096	16095	16093	16092	16090	16089
.72	16087	16086	16084	16083	16081	16079	16077	16076	16074	16073
.73	16071	16070	16068	16067	16065	16064	16062	16061	16059	16058
.74	16056	16055	16053	16052	16050	16049	16047	16046	16044	16043
16.75	16041	16040	16038	16037	16035	16033	16031	16030	16028	16027
.76	16025	16024	16022	16021	16019	16017	16015	16014	16012	16011
.77	16009	16008	16006	16005	16003	16002	16000	15999	15997	15996
.78	15994	15993	15991	15990	15988	15986	15984	15983	15981	15980
.79	15978	15977	15975	15974	15972	15971	15969	15968	15966	15965
.80	15963	15962	15960	15959	15957	15955	15953	15952	15950	15949
.81	15947	15946	15944	15943	15941	15939	15937	15936	15934	15933
.82	15931	15930	15928	15927	15925	15924	15922	15921	15919	15918
.83	15916	15915	15913	15912	15910	15909	15907	15906	15904	15903
.84	15901	15900	15898	15897	15895	15893	15891	15890	15888	15887
.85	15885	15884	15882	15881	15879	15878	15876	15875	15873	15872
.86	15870	15869	15867	15866	15864	15862	15860	15859	15857	15856
.87	15854	15853	15851	15850	15848	15847	15845	15844	15842	15841
.88	15839	15838	15836	15835	15833	15831	15829	15828	15826	15825
.89	15823	15822	15820	15819	15817	15816	15814	15813	15811	15810
.90	15808	15807	15805	15804	15802	15801	15799	15798	15796	15795
.91	15793	15792	15790	15789	15787	15785	15783	15782	15780	15779
.92	15777	15776	15774	15773	15771	15770	15768	15767	15765	15764
.93	15762	15761	15759	15758	15756	15755	15753	15752	15750	15749
.94	15747	15746	15744	15743	15741	15739	15737	15736	15734	15733
.95	15731	15730	15728	15727	15725	15723	15721	15720	15718	15717
.96	15715	15714	15712	15711	15709	15708	15706	15705	15703	15702
.97	15700	15699	15697	15696	15694	15693	15691	15690	15688	15687
.98	15685	15684	15682	15681	15679	15678	15676	15675	15673	15672
16.99	15670	15669	15667	15666	15664	15663	15661	15660	15658	15657
In.dec.	0	1	2	3	4	5	6	7	8	9

In.dec.	In. dec. 17.000					In. dec. 17.499				
In.dec.	0	1	2	3	4	5	6	7	8	9
17.00	15655	15654	15652	15651	15649	15648	15646	15645	15643	15642
.01	15640	15639	15637	15636	15634	15632	15630	15629	15627	15626
.02	15624	15623	15621	15620	15618	15617	15615	15614	15612	15611
.03	15609	15608	15606	15605	15603	15601	15599	15598	15596	15595
.04	15593	15592	15590	15589	15587	15586	15584	15583	15581	15580
.05	15578	15577	15575	15574	15572	15571	15569	15568	15566	15565
.06	15563	15562	15560	15559	15557	15556	15554	15553	15551	15550
.07	15548	15547	15545	15544	15542	15540	15538	15537	15535	15534
.08	15532	15531	15529	15528	15526	15525	15523	15522	15520	15519
.09	15517	15516	15514	15513	15511	15510	15508	15507	15505	15504
.10	15502	15501	15499	15498	15496	15495	15493	15492	15490	15489
.11	15487	15486	15484	15483	15481	15480	15478	15477	15475	15474
.12	15472	15471	15469	15468	15466	15464	15462	15461	15459	15458
.13	15456	15455	15453	15452	15450	15449	15447	15446	15444	15443
.14	15441	15440	15438	15437	15435	15434	15432	15431	15429	15428
.15	15426	15425	15423	15422	15420	15418	15416	15415	15413	15412
.16	15410	15409	15407	15406	15404	15403	15401	15400	15398	15397
.17	15395	15394	15392	15391	15389	15388	15386	15385	15383	15382
.18	15380	15379	15377	15376	15534	15373	15371	15370	15368	15367
.19	15365	15364	15362	15361	15359	15358	15356	15355	15353	15352
.20	15350	15349	15347	15346	15344	15343	15341	15340	15338	15337
.21	15335	15334	15332	15331	15329	15328	15326	15325	15323	15322
.22	15320	15319	15317	15316	15314	15312	15310	15309	15307	15306
.23	15304	15304	15302	15301	15299	15297	15295	15294	15292	15291
.24	15289	15288	15286	15285	15283	15282	15280	15279	15277	15276
17.25	15274	15273	15271	15270	15268	15267	15265	15264	15262	15261
.26	15259	15258	15256	15255	15253	15252	15250	15249	15247	15246
.27	15244	15243	15241	15240	15238	15237	15235	15234	15232	15231
.28	15229	15228	15226	15225	15223	15222	15220	15219	15217	15216
.29	15214	15213	15211	15210	15208	15207	15205	15204	15202	15201
.30	15199	15198	15196	15195	15193	15192	15190	15189	15187	15186
.31	15184	15183	15181	15180	15178	15177	15175	15174	15172	15171
.32	15169	15168	15166	15165	17163	15162	15160	15159	15157	15156
.33	15154	15153	15151	15150	15148	15147	15145	15144	15142	15141
.34	15139	15138	15136	15135	15133	15132	15130	15129	15127	15126
.35	15124	15123	15121	15120	15118	15117	15115	15114	15112	15111
.36	15109	15108	15106	15105	15103	15102	15100	15099	15097	15096
.37	15094	15093	15091	15090	15088	15087	15085	15084	15082	15081
.38	15079	15078	15076	15075	15073	15072	15070	15069	15067	15066
.39	15064	15063	15061	15060	15058	15057	15055	15054	15052	15051
.40	15049	15048	15046	15045	15043	15042	15040	15039	15037	15036
.41	15034	15033	15031	15030	15028	15027	15025	15024	15022	15021
.42	15019	15018	15016	15015	15013	15012	15010	15009	15007	15006
.43	15004	15003	15001	15000	14998	14997	14995	14994	14992	14991
.44	14989	14988	14986	14985	14983	14982	14980	14979	14977	14976
.45	14974	14973	14971	14970	14968	14967	14965	14964	14962	14961
.46	14959	14958	14956	14955	14953	14952	14950	14949	14947	14946
.47	14944	14943	14941	14940	14938	14937	14935	14934	14932	14931
.48	14929	14928	14926	14925	14923	14922	14920	14919	14917	14916
17.49	14914	14913	14911	14910	14908	14907	14905	14904	14902	14901
In.dec.	0	1	2	3	4	5	6	7	8	9

In.dec.	In. dec. 17.500					In. dec. 17.999				
In.dec.	0	1	2	3	4	5	6	7	8	9
17.50	14899	14898	14896	14895	14893	14892	14890	14889	14887	14886
.51	14884	14883	14881	14880	14878	14877	14875	14874	14872	14871
.52	14870	14869	14867	14866	14864	14863	14861	14860	14858	14857
.53	14855	14854	14852	14851	14849	14848	14846	14845	14843	14842
.54	14840	14839	14837	14836	14834	14833	14831	14830	14828	14827
.55	14825	14824	14822	14821	14819	14818	14816	14815	14813	14812
.56	14810	14809	14807	14806	14804	14803	14801	14800	14798	14797
.57	14795	14794	14792	14791	14789	14788	14786	14785	14783	14782
.58	14780	14779	14777	14776	14774	14773	14771	14770	14768	14767
.59	14765	14764	14762	14761	14759	14758	14756	14755	14753	14752
.60	14751	14750	14748	14747	14745	14744	14742	14741	14739	14738
.61	14736	14735	14733	14732	14730	14729	14727	14726	14724	14723
.62	14721	14720	14718	14717	14715	14714	14712	14711	14709	14708
.63	14707	14706	14704	14703	14701	14700	14698	14697	14695	14694
.64	14692	14691	14689	14688	14686	14685	14683	14682	14680	14679
.65	14677	14676	14674	14673	14671	14670	14668	14667	14665	14664
.66	14662	14661	14659	14658	14656	14655	14653	14652	14650	14649
.67	14647	14646	14644	14643	14641	14640	14638	14637	14635	14634
.68	14633	14632	14630	14629	14627	14626	14624	14623	14621	14620
.69	14618	14617	14615	14614	14612	14611	14609	14608	14606	14605
.70	14603	14602	14600	14599	14597	14596	14594	14593	14591	14590
.71	14588	14587	14585	14584	14582	14581	14579	14578	14576	14575
.72	14574	14573	14571	14570	14568	14567	14565	14564	14562	14561
.73	14559	14558	14556	14555	14553	14552	14550	14549	14547	14546
.74	14545	14544	14542	14541	14539	14538	14536	14535	14533	14532
17.75	14530	14529	14527	14526	14524	14523	14521	14520	14518	14517
76	14515	14514	14512	14511	14509	14508	14506	14505	14503	14502
77	14500	14499	14497	14496	14494	14493	14491	14490	14488	14487
.78	14486	14485	14483	14482	14480	14479	14477	14476	14474	14473
79	14471	14470	14468	14467	14465	14464	14462	14461	14459	14458
.80	14456	14455	14453	14452	14450	14449	14447	14446	14444	14443
.81	14442	14441	14439	14438	14436	14435	14433	14432	14430	14429
.82	14427	14426	14424	14423	14421	14420	14418	14417	14415	14414
.83	14413	14412	14410	14409	14407	14406	14404	14403	14401	14400
.84	14398	14397	14395	14394	14392	14391	14389	14388	14386	14385
.85	14383	14382	14380	14379	14377	14376	14374	14373	14371	14370
.86	14369	14368	14366	14365	14363	14362	14360	14359	14357	14356
.87	14354	14353	14351	14350	14348	14347	14345	14344	14342	14341
.88	14339	14338	14336	14335	14333	14332	14330	14329	14327	14326
.89	14325	14324	14322	14321	14319	14318	14316	14315	14313	14312
.90	14311	14310	14308	14307	14305	14304	14302	14301	14299	14298
.91	14296	14295	14293	14292	14290	14289	14287	14286	14284	14283
.92	14281	14280	14278	14277	14275	14274	14272	14271	14269	14268
.93	14267	14266	14264	14263	14261	14260	14258	14257	14255	14254
.94	14252	14251	14249	14248	14246	14245	14243	14242	14240	14239
.95	14238	14237	14235	14234	14232	14231	14229	14228	14226	14225
.96	14223	14222	14220	14219	14217	14216	14214	14213	14211	14210
.97	14209	14208	14206	14205	14203	14202	14200	14199	14197	14196
.98	14194	14193	14191	14190	14188	14187	14185	14184	14182	14181
17.99	14180	14179	14177	14176	14174	14173	14171	14170	14168	14167
In.dec.	0	1	2	3	4	5	6	7	8	9

	In. dec. 18.000					In. dec. 18 499				
In.dec.	0	1	2	3	4	5	6	7	8	9
18. 00	14165	14163	14162	14161	14159	14158	14156	14155	14153	14152
.01	14151	14150	14148	14147	14145	14144	14142	14141	14139	14138
.02	14137	14136	14134	14133	14131	14130	14128	14127	14125	14124
.03	14122	14121	14119	14118	14116	14115	14113	14112	14110	14109
.04	14107	14106	14104	14103	14101	14100	14098	14097	14095	14094
.05	14093	14092	14090	14089	14087	14086	14084	14083	14081	14080
.06	14078	14077	14075	14074	14072	14071	14069	14068	14066	14065
.07	14064	14063	14061	14060	14058	14057	14055	14054	14052	14051
.08	14050	14049	14047	14046	14044	14043	14041	14040	14038	14037
.09	14035	14034	14032	14031	14029	14028	14026	14025	14023	14022
.10	14021	14020	14018	14017	14015	14014	14012	14011	14009	14008
.11	14006	14005	14003	14002	14000	13999	13997	13996	13994	13993
.12	13992	13991	13989	13988	13986	13985	13983	13982	13980	13979
13	13978	13977	13975	13974	13972	13971	13969	13968	13966	13965
.14	13963	13962	13960	13959	13957	13956	13954	13953	13951	13950
.15	13949	13948	13946	13945	13943	13942	13940	13939	13937	13936
.16	13934	13933	13931	13930	13928	13927	13925	13924	13922	13921
.17	13920	13919	13917	13916	13914	13913	13911	13910	13908	13907
.18	13906	13905	13903	13902	13900	13899	13897	13896	13894	13893
.19	13892	13891	13889	13888	13886	13885	13883	13882	13880	13879
.20	13877	13876	13874	13873	13871	13870	13868	13867	13865	13864
.21	13863	13862	13860	13859	13857	13856	13854	13853	13851	13850
.22	13849	13848	13846	13845	13843	13842	13840	13839	13837	13836
.23	13834	13833	13831	13830	13828	13827	13825	13824	13822	13821
.24	13820	13819	13817	13816	13814	13813	13811	13810	13808	13807
18. 25	13806	13805	13803	13802	13800	13799	13797	13796	13794	13793
.26	13792	13791	13789	13788	13786	13785	13783	13782	13780	13779
.27	13777	13776	13774	13773	13771	13770	13768	13767	13765	13764
.28	13763	13762	13760	13759	13757	13756	13754	13753	13751	13750
.29	13749	13748	13746	13745	13743	13742	13740	13739	13737	13736
.30	13734	13733	13731	13730	13728	13727	13725	13724	13722	13721
.31	13720	13719	13717	13716	13714	13713	13711	13710	13708	13707
.32	13706	13705	13703	13702	13700	13699	13697	13696	13694	13693
.33	13692	13691	13689	13688	13686	13685	13683	13682	13680	13679
.34	13678	13677	13675	13674	13672	13671	13669	13668	13666	13665
.35	13663	13662	13660	13659	13657	13656	13654	13653	13651	13650
.36	13649	13648	13646	13645	13643	13642	13640	13639	13637	13636
.37	13635	13634	13632	13631	13629	13628	13626	13625	13623	13622
.38	13621	13620	13618	13617	13615	13614	13612	13611	13609	13608
.39	13607	13606	13604	13603	13601	13600	13598	13597	13595	13594
.40	13592	13591	13589	13588	13586	13585	13583	13582	13580	13579
.41	13578	13577	13575	13574	13572	13571	13569	13568	13566	13565
.42	13564	13563	13561	13560	13558	13557	13555	13554	13552	13551
.43	13550	13549	13547	13546	13544	13543	13541	13540	13538	13537
.44	13536	13535	13533	13532	13530	13529	13527	13526	13524	13523
.45	13522	13521	13519	13518	13516	13515	13513	13512	13510	13509
.46	13508	13507	13505	13504	13502	13501	13499	13498	13496	13495
.47	13493	13492	13490	13489	13487	13486	13484	13483	13481	13480
.48	13479	13478	13476	13475	13473	13472	13470	13469	13467	13466
18. 49	13465	13464	13462	13461	13459	13458	13456	13455	13453	13452
In.dec.	0	1	2	3	4	5	6	7	8	9

In.dec	In. dec. 18.500					In. dec. 18.999				
	0	1	2	3	4	5	6	7	8	9
18.50	13451	13450	13448	13447	13445	13444	13442	13441	13439	13438
.51	13437	13436	13434	13433	13431	13430	13428	13427	13425	13424
.52	13423	13422	13420	13419	13417	13416	13414	13413	13411	13410
.53	13409	13408	13406	13405	13403	13402	13400	13399	13397	13396
.54	13395	13394	13392	13391	13389	13388	13386	13385	13383	13382
.55	13381	13380	13378	13377	13375	13374	13372	13371	13369	13368
.56	13367	13366	13364	13363	13361	13360	13358	13357	13355	13354
.57	13353	13352	13350	13349	13347	13346	13344	13343	13341	13340
.58	13339	13338	13336	13335	13333	13332	13330	13329	13327	13326
.59	13325	13324	13322	13321	13319	13318	13316	13315	13313	13312
.60	13311	13310	13308	13307	13305	13304	13302	13301	13299	13298
.61	13297	13296	13294	13293	13291	13290	13288	13287	13285	13284
.62	13283	13282	13280	13279	13277	13276	13274	13273	13271	13270
.63	13269	13268	13266	13265	13263	13262	13260	13259	13257	13256
.64	13255	13254	13252	13251	13249	13248	13246	13245	13243	13242
.65	13241	13240	13238	13237	13235	13234	13232	13231	13229	13228
.66	13227	13226	13224	13223	13221	13220	13218	13217	13215	13214
.67	13213	13212	13210	13209	13207	13206	13204	13203	13201	13200
.68	13199	13198	13196	13195	13193	13192	13190	13189	13187	13186
.69	13185	13184	13182	13181	13179	13178	13176	13175	13173	13172
.70	13171	13170	13168	13167	13165	13164	13162	13161	13159	13158
71	13157	13156	13154	13153	13151	13150	13148	13147	13145	13144
72	13143	13142	13140	13139	13137	13136	13134	13133	13131	13130
.73	13129	13128	13126	13125	13123	13122	13120	13119	13117	13116
74	13115	13114	13113	13111	13110	13108	13107	13105	13104	13103
18.75	13102	13101	13099	13098	13096	13095	13093	13092	13090	13089
76	13088	13087	13085	13084	13082	13081	13079	13078	13076	13075
77	13074	13073	13071	13070	13068	13067	13065	13064	13062	13061
.78	13060	13059	13057	13056	13054	13053	13051	13050	13048	13047
.79	13046	13045	13043	13042	13040	13039	13037	13036	13034	13033
.80	13032	13031	13029	13028	13026	13025	13023	13022	13020	13019
.81	13018	13017	13015	13014	13012	13011	13009	13008	13006	13005
.82	13004	13003	13001	13000	12998	12997	12995	12994	12992	12991
.83	12990	12989	12988	12986	12985	12983	12982	12980	12979	12978
.84	12977	12976	12974	12973	12971	12970	12968	12967	12965	12964
.85	12963	12962	12960	12959	12957	12956	12954	12953	12951	12950
.86	12949	12948	12946	12945	12943	12942	12940	12939	12937	12936
.87	12935	12934	12933	12931	12930	12928	12927	12925	12924	12923
.88	12922	12921	12919	12918	12916	12915	12913	12912	12910	12909
.89	12908	12907	12905	12904	12902	12901	12899	12898	12896	12895
.90	12894	12893	12891	12890	12888	12887	12885	12884	12882	12881
.91	12880	12879	12877	12876	12874	12873	12871	12870	12868	12867
.92	12866	12865	12863	12862	12860	12859	12857	12856	12854	12853
.93	12852	12851	12850	12848	12847	12845	12844	12842	12841	12840
.94	12839	12838	12836	12835	12833	12832	12830	12829	12827	12826
.95	12825	12824	12822	12821	12819	12818	12816	12815	12813	12812
.96	12811	12810	12808	12807	12805	12804	12802	12801	12799	12798
.97	12797	12796	12795	12793	12792	12790	12789	12787	12786	12785
.98	12784	12783	12781	12780	12778	12777	12775	12774	12772	12771
18.99	12770	12769	12768	12766	12765	12763	12762	12760	12759	12758
In.dec.	0	1	2	3	4	5	6	7	8	9

	In. dec. 19.000					In. dec. 19.499				
ln.dec	0	1	2	3	4	5	6	7	8	9
19.00	12757	12756	12754	12753	12751	12750	12748	12747	12745	12744
.01	12743	12742	12740	12739	12737	12736	12734	12733	12731	12730
.02	12729	12728	12726	12725	12723	12722	12720	12719	12717	12716
.03	12715	12714	12712	12711	12709	12708	12706	12705	12703	12702
.04	12701	12700	12699	12697	12696	12694	12693	12691	12690	12689
.05	12688	12687	12685	12684	12682	12681	12679	12678	12676	12675
.06	12674	12673	12672	12670	12669	12667	12666	12664	12663	12662
.07	12661	12660	12658	12657	12655	12654	12652	12651	12649	12648
.08	12647	12646	12644	12643	12641	12640	12638	12637	12635	12634
.09	12633	12632	12631	12629	12628	12626	12625	12623	12622	12621
.10	12620	12619	12617	12616	12614	12613	12611	12610	12608	12607
.11	12606	12605	12603	12602	12600	12599	12597	12596	12594	12593
.12	12592	12591	12590	12588	12587	12585	12584	12582	12581	12580
.13	12579	12578	12576	12575	12573	12572	12570	12569	12567	12566
.14	12565	12564	12562	12561	12559	12558	12556	12555	12553	12552
.15	12551	12550	12549	12547	12546	12544	12543	12541	12540	12539
.16	12538	12537	12535	12534	12532	12531	12529	12528	12526	12525
.17	12524	12523	12522	12520	12519	12517	12516	12514	12513	12512
.18	12511	12510	12508	12507	12505	12504	12502	12501	12499	12498
.19	12497	12496	12495	12493	12492	12490	12489	12487	12486	12485
.20	12484	12483	12481	12480	12478	12477	12475	12474	12472	12471
.21	12470	12469	12468	12466	12465	12463	12462	12460	12459	12458
.22	12457	12456	12454	12453	12451	12450	12448	12447	12445	12444
.23	12443	12442	12440	12439	12437	12436	12434	12433	12431	12430
.24	12429	12428	12427	12425	12424	12422	12421	12419	12418	12417
19.25	12416	12415	12413	12412	12410	12409	12407	12406	12404	12403
.26	12402	12401	12400	12398	12397	12395	12394	12392	12391	12390
.27	12389	12388	12386	12385	12383	12382	12380	12379	12377	12376
.28	12375	12374	12373	12371	12370	12368	12367	12365	12364	12363
.29	12362	12361	12359	12358	12356	12355	12353	12352	12350	12349
.30	12348	12347	12346	12344	12343	12341	12340	12338	12337	12336
.31	12335	12334	12332	12331	12329	12328	12326	12325	12323	12322
.32	12321	12320	12319	12317	12316	12314	12313	12311	12310	12309
.33	12308	12307	12305	12304	12302	12301	12299	12298	12296	12295
.34	12294	12293	12292	12290	12289	12287	12286	12284	12283	12282
.35	12281	12280	12278	12277	12275	12274	12272	12271	12269	12268
.36	12267	12266	12265	12263	12262	12260	12259	12257	12256	12255
.37	12254	12253	12252	12250	12249	12247	12246	12244	12243	12242
.38	12241	12240	12238	12237	12235	12234	12232	12231	12229	12228
.39	12227	12226	12225	12223	12222	12220	12219	12217	12216	12215
.40	12214	12213	12211	12210	12208	12207	12205	12204	12202	12201
.41	12200	12199	12198	12196	12195	12193	12192	12190	12189	12188
.42	12187	12186	12184	12183	12181	12180	12178	12177	12175	12174
.43	12173	12172	12171	12169	12168	12166	12165	12163	12162	12161
.44	12160	12159	12157	12156	12154	12153	12151	12150	12148	12147
.45	12146	12145	12144	12142	12141	12139	12138	12136	12135	12134
.46	12133	12132	12130	12129	12127	12126	12124	12123	12121	12120
.47	12119	12118	12117	12115	12114	12112	12111	12109	12108	12107
.48	12106	12105	12104	12102	12101	12099	12098	12096	12095	12094
19.49	12093	12092	12091	12089	12088	12086	12085	12083	12082	12081
ln.dec.	0	1 +	2	3	4	5	6	7	8	9

ln. dec.	ln. dec. 19 500					ln. dec. 19.999			
ln.dec.	**0**	**1**	**2**	**3**	**4**	**5**	**6**	**7**	**8**	**9**
19.50	12080	12079	12077	12076	12074	12073	12071	12070	12068	12067
.51	12066	12065	12064	12062	12061	12059	12058	12056	12055	12054
.52	12053	12052	12050	12049	12047	12046	12044	12043	12041	12040
.53	12039	12038	12037	12035	12034	12032	12031	12029	12028	12027
.54	12026	12025	12024	12022	12021	12019	12018	12016	12015	12014
.55	12013	12012	12010	12009	12007	12006	12004	12003	12001	12000
.56	11999	11998	11997	11995	11994	11992	11991	11989	11988	11987
.57	11986	11985	11984	11982	11981	11979	11978	11976	11975	11974
.58	11973	11972	11971	11969	11968	11966	11965	11963	11962	11961
.59	11960	11959	11957	11956	11954	11953	11951	11950	11948	11947
.60	11946	11945	11944	11942	11941	11939	11938	11936	11935	11934
.61	11933	11932	11931	11929	11928	11926	11925	11923	11922	11921
.62	11920	11919	11917	11916	11914	11913	11911	11910	11908	11907
.63	11906	11905	11904	11902	11901	11899	11898	11896	11895	11894
.64	11893	11892	11891	11889	11888	11886	11885	11883	11882	11881
.65	11880	11879	11878	11876	11875	11873	11872	11870	11869	11868
.66	11867	11866	11865	11863	11862	11860	11859	11857	11856	11855
.67	11854	11853	11851	11850	11848	11847	11845	11844	11842	11841
.68	11840	11839	11838	11836	11835	11833	11832	11830	11829	11828
.69	11827	11826	11824	11823	11821	11820	11818	11817	11815	11814
70	11813	11812	11811	11809	11808	11806	11805	11803	11802	11801
71	11800	11799	11798	11796	11795	11793	11792	11790	11789	11788
72	11787	11786	11785	11783	11782	11780	11779	11777	11776	11775
73	11774	11773	11771	11770	11768	11767	11765	11764	11762	11761
74	11760	11759	11758	11756	11755	11753	11752	11750	11749	11748
19.75	11747	11746	11745	11743	11742	11740	11739	11737	11736	11735
76	11734	11733	11732	11730	11729	11727	11726	11724	11723	11722
77	11721	11720	11719	11717	11716	11714	11713	11711	11710	11709
78	11708	11707	11705	11704	11702	11701	11699	11698	11696	11695
.79	11694	11693	11692	11690	11689	11687	11686	11684	11683	11682
.80	11681	11680	11679	11678	11676	11675	11673	11672	11671	11670
.81	11669	11668	11667	11665	11664	11662	11661	11659	11658	11657
.82	11656	11655	11653	11652	11650	11649	11647	11646	11644	11643
.83	11642	11641	11640	11638	11637	11635	11634	11632	11631	11630
.84	11629	11628	11627	11625	11624	11622	11621	11619	11618	11617
.85	11616	11615	11614	11612	11611	11609	11608	11606	11605	11604
.86	11603	11602	11601	11599	11598	11596	11595	11593	11592	11591
.87	11590	11589	11588	11586	11585	11583	11582	11580	11579	11578
.88	11577	11576	11575	11573	11572	11570	11569	11567	11566	11565
.89	11564	11563	11562	11560	11559	11557	11556	11554	11553	11552
.90	11551	11550	11549	11547	11546	11544	11543	11541	11540	11539
.91	11538	11537	11535	11534	11532	11531	11529	11528	11526	11525
.92	11524	11523	11522	11520	11519	11517	11516	11514	11513	11512
.93	11511	11510	11509	11507	11506	11504	11503	11501	11500	11499
.94	11498	11497	11496	11494	11493	11491	11490	11488	11487	11486
.95	11485	11484	11483	11482	11480	11479	11477	11476	11475	11474
.96	11473	11472	11470	11469	11467	11466	11464	11463	11461	11460
.97	11459	11458	11457	11455	11454	11452	11451	11449	11448	11447
.98	11446	11445	11444	11442	11441	11439	11438	11436	11435	11434
19.99	11433	11432	11431	11429	11428	11426	11425	11423	11422	11421
ln. dec.	**0**	**1**	**2**	**3**	**4**	**5**	**6**	**7**	**8**	**9**

	In. dec. 20.000					In. dec. 20.499				
In.dec	0	1	2	3	4	5	6	7	8	9
20.00	11420	11419	11418	11416	11415	11413	11412	11410	11409	11408
.01	11407	11406	11405	11403	11402	11400	11399	11397	11396	11395
.02	11394	11393	11392	11390	11389	11387	11386	11384	11383	11382
.03	11381	11380	11379	11377	11376	11374	11373	11371	11370	11369
.04	11368	11367	11366	11364	11363	11361	11360	11358	11357	11356
.05	11355	11354	11353	11351	11350	11348	11347	11345	11344	11343
.06	11342	11341	11340	11338	11337	11335	11334	11332	11331	11330
.07	11329	11328	11327	11325	11324	11322	11321	11319	11318	11317
.08	11316	11315	11314	11312	11311	11309	11308	11306	11305	11304
.09	11303	11302	11301	11299	11298	11296	11295	11293	11292	11291
.10	11290	11289	11288	11286	11285	11283	11282	11280	11279	11278
.11	11277	11276	11275	11273	11272	11270	11269	11267	11266	11265
.12	11264	11263	11262	11260	11259	11257	11256	11254	11253	11252
.13	11251	11250	11249	11247	11246	11244	11243	11241	11240	11239
.14	11238	11237	11236	11234	11233	11231	11230	11228	11227	11226
.15	11225	11224	11223	11221	11220	11218	11217	11215	11214	11213
.16	11212	11211	11210	11208	11207	11205	11204	11202	11201	11200
.17	11199	11198	11197	11195	11194	11192	11191	11189	11188	11187
.18	11186	11185	11184	11182	11181	11179	11178	11176	11175	11174
.19	11173	11172	11171	11169	11168	11166	11165	11163	11162	11161
.20	11160	11159	11158	11156	11155	11153	11152	11150	11149	11148
.21	11147	11146	11145	11144	11142	11141	11139	11138	11137	11136
.22	11135	11134	11133	11131	11130	11128	11127	11125	11124	11123
.23	11122	11121	11120	11118	11117	11115	11114	11112	11111	11110
.24	11109	11108	11107	11105	11104	11102	11101	11099	11098	11097
20.25	11096	11095	11094	11092	11091	11089	11088	11086	11085	11084
.26	11083	11082	11081	11080	11079	11077	11075	11073	11072	11071
.27	11070	11069	11068	11066	11065	11063	11062	11060	11059	11058
.28	11057	11056	11055	11054	11052	11051	11049	11048	11047	11046
.29	11045	11044	11043	11041	11040	11038	11037	11035	11034	11033
.30	11032	11031	11030	11028	11027	11025	11024	11022	11021	11020
.31	11019	11018	11017	11015	11014	11012	11011	11009	11008	11007
.32	11006	11005	11004	11002	11001	10999	10998	10996	10995	10994
.33	10993	10992	10991	10990	10988	10987	10985	10984	10983	10982
.34	10981	10980	10979	10977	10976	10974	10973	10971	10970	10969
.35	10968	10967	10966	10964	10963	10961	10960	10958	10957	10956
.36	10955	10954	10953	10951	10950	10948	10947	10945	10944	10943
.37	10942	10941	10940	10938	10937	10935	10934	10932	10931	10930
.38	10929	10928	10927	10925	10924	10922	10921	10919	10918	10917
.39	10916	10915	10914	10913	10911	16910	10908	10907	10906	10905
.40	10904	10903	10902	10900	10899	10897	10896	10894	10893	10892
.41	10891	10890	10889	10887	10886	10884	10883	10881	10880	10879
.42	10878	10877	10876	10874	10873	10871	10870	10868	10867	10666
.43	10865	10864	10863	10862	10860	10859	10857	10856	10855	10854
.44	10853	10852	10851	10849	10848	10846	10845	10843	10842	10841
.45	10840	10839	10838	10836	10835	10833	10832	10830	10829	10828
.46	10827	10826	10825	10823	10822	10820	10819	10817	10816	10815
.47	10814	10813	10812	10811	10809	10808	10806	10805	10804	10803
.48	10802	10801	10800	10798	10797	10795	10794	10792	10791	10790
20.49	10789	10788	10787	10785	10784	10782	10781	10779	10778	10777
In.dec	0	1	2	3	4	5	6	7	8	9

	In. dec. 20.500					In. dec. 20.999				
In.dec.	0	1	2	3	4	5	6	7	8	9
20.50	10776	10775	10774	10772	10771	10769	10768	10766	10765	10764
.51	10763	10762	10761	10760	10758	10757	10755	10754	10753	10752
.52	10751	10750	10749	10747	10746	10744	10743	10741	10740	10739
.53	10738	10737	10736	10734	10733	10731	10730	10728	10727	10726
.54	10725	10724	10723	10722	10720	10719	10717	10716	10715	10714
.55	10713	10712	10711	10709	10708	10705	10705	10703	10702	10701
.56	10700	10699	10698	10697	10695	10694	10692	10691	10690	10689
.57	10688	10687	10686	10684	10683	10681	10680	10678	10677	10676
.58	10675	10674	10673	10671	10670	10668	10667	10665	10664	10663
.59	10662	10661	10660	10658	10657	10655	10654	10652	10651	10650
.60	10649	10648	10647	10646	10644	10643	10641	10640	10639	10638
.61	10637	10636	10635	10633	10632	10630	10629	10627	10626	10625
.62	10624	10623	10622	10621	10619	10618	10616	10615	10614	10613
.63	10612	10611	10610	10608	10607	10605	10604	10602	10601	10600
.64	10599	10598	10597	10595	10594	10592	10591	10589	10588	10587
.65	10586	10585	10584	10582	10581	10579	10578	10576	10575	10574
.66	10573	10572	10571	10570	10568	10567	10565	10564	10563	10562
.67	10561	10560	10559	10558	10556	10555	10553	10552	10551	10550
.68	10549	10548	10547	10545	10544	10542	10541	10539	10538	10537
.69	10536	10535	10534	10532	10531	10529	10528	10526	10525	10524
.70	10523	10522	10521	10520	10518	10517	10515	10514	10513	10512
.71	10511	10510	10509	10507	10506	10504	10503	10501	10500	10499
.72	10498	10497	10496	10495	10493	10492	10490	10489	10488	10487
73	10486	10485	10484	10482	10481	10479	10478	10476	10475	10474
.74	10473	10472	10471	10469	10468	10466	10465	10463	10462	10461
20.75	10460	10459	10458	10457	10455	10454	10452	10451	10450	10449
.76	10448	10447	10446	10444	10443	10441	10440	10438	10437	10436
.77	10435	10434	10433	10432	10430	10429	10427	10426	10425	10424
.78	10423	10422	10421	10420	10418	10417	10415	10414	10413	10412
.79	10411	10410	10409	10407	10406	10404	10403	10401	10400	10399
.80	10398	10397	10396	10394	10393	10391	10390	10388	10387	10386
.81	10385	10384	10383	10382	10380	10379	10377	10376	10375	10374
.82	10373	10372	10371	10369	10368	10366	10365	10363	10362	10361
.83	10360	10359	10358	10357	10355	10354	10352	10351	10350	10349
.84	10348	10347	10346	10344	10343	10341	10340	10338	10337	10336
.85	10335	10334	10333	10332	10330	10329	10327	10326	10325	10324
.86	10323	10322	10321	10319	10318	10316	10315	10313	10312	10311
.87	10310	10309	10308	10307	10305	10304	10302	10301	10300	10299
.88	10298	10297	10296	10294	10293	10291	10290	10288	10287	10286
.89	10285	10284	10283	10282	10280	10279	10277	10276	10275	10274
.90	10273	10272	10271	10270	10268	10267	10265	10264	10263	10262
.91	10261	10260	10259	10257	10256	10254	10253	10251	10250	10249
.92	10248	10247	10246	10244	10243	10241	10240	10238	10237	10236
.93	10235	10234	10233	10232	10230	10229	10227	10226	10225	10224
.94	10223	10222	10221	10220	10218	10217	10215	10214	10213	10212
.95	10211	10210	10209	10207	10206	10204	10203	10201	10200	10199
.96	10198	10197	10196	10195	10193	10192	10190	10189	10188	10187
.97	10186	10185	10184	10182	10181	10179	10178	10176	10175	10174
.98	10173	10172	10171	10170	10168	10167	10165	10164	10163	10162
20.99	10161	10160	10159	10157	10156	10154	10153	10151	10150	10149
In.dec.	0	1	2	3	4	5	6	7	8	9

In.dec.	In. dec. 21.000					In. dec. 21.499				
	0	1	2	3	4	5	6	7	8	9
21.00	10148	10147	10146	10145	10143	10142	10140	10139	10138	10137
.01	10136	10135	10134	10133	10131	10130	10128	10127	10126	10125
.02	10124	10123	10122	10120	10119	10117	10116	10114	10113	10112
.03	10111	10110	10109	10108	10106	10105	10103	10102	10101	10100
.04	10099	10098	10097	10096	10094	10093	10091	10090	10089	10088
.05	10087	10086	10085	10083	10082	10080	10079	10077	10076	10075
.06	10074	10073	10072	10071	10069	10068	10066	10065	10064	10063
.07	10062	10061	10060	10058	10057	10055	10054	10052	10051	10050
.08	10049	10048	10047	10046	10044	10043	10041	10040	10039	10038
.09	10037	10036	10035	10034	10032	10031	10029	10028	10027	10026
.10	10025	10024	10023	10021	10020	10018	10017	10015	10014	10013
.11	10012	10011	10010	10009	10008	10006	10004	10003	10002	10001
.12	10000	9999	9998	9997	9995	9994	9992	9991	9990	9989
.13	9988	9987	9986	9985	9983	9982	9980	9979	9978	9977
.14	9976	9975	9974	9972	9971	9969	9968	9966	9965	9964
.15	9963	9962	9961	9959	9958	9956	9955	9953	9952	9951
.16	9950	9949	9948	9947	9945	9944	9942	9941	9940	9939
.17	9938	9937	9936	9935	9933	9932	9930	9929	9928	9927
.18	9926	9925	9924	9923	9921	9920	9918	9917	9916	9915
.19	9914	9913	9912	9910	9909	9907	9906	9904	9903	9902
.20	9901	9900	9899	9898	9896	9895	9893	9892	9891	9890
.21	9889	9888	9887	9886	9884	9883	9881	9880	9879	9878
.22	9877	9876	9875	9874	9872	9871	9869	9868	9867	9866
.23	9865	9864	9863	9862	9860	9859	9857	9856	9855	9854
.24	9853	9852	9851	9849	9848	9846	9845	9843	9842	9841
21.25	9840	9839	9838	9837	9835	9834	9832	9831	9830	9829
.26	9828	9827	9826	9824	9823	9821	9820	9818	9817	9816
.27	9815	9814	9813	9812	9810	9809	9807	9806	9805	9804
.28	9803	9802	9801	9800	9798	9797	9795	9794	9793	9792
.29	9791	9790	9789	9788	9786	9785	9783	9782	9781	9780
.30	9779	9778	9777	9776	9774	9773	9771	9770	9769	9768
.31	9767	9766	9765	9763	9762	9760	9759	9757	9756	9755
.32	9754	9753	9752	9751	9749	9748	9746	9745	9744	9743
.33	9742	9741	9740	9739	9737	9736	9734	9733	9732	9731
.34	9730	9729	9728	9727	9725	9724	9722	9721	9720	9719
.35	9718	9717	9716	9715	9713	9712	9710	9709	9708	9707
.36	9706	9705	9704	9703	9701	9700	9698	9697	9696	9695
.37	9694	9693	9692	9690	9689	9687	9686	9684	9683	9682
.38	9681	9680	9679	9678	9676	9675	9673	9672	9671	9670
.39	9669	9668	9667	9666	9664	9663	9661	9660	9659	9658
.40	9657	9656	9655	9653	9652	9650	9649	9647	9646	9645
.41	9644	9643	9642	9641	9639	9638	9636	9635	9634	9633
.42	9632	9631	9630	9629	9627	9626	9624	9623	9622	9621
.43	9620	9619	9618	9617	9615	9614	9612	9611	9610	9609
.44	9608	9607	9606	9605	9603	9602	9600	9599	9598	9597
.45	9596	9595	9594	9593	9591	9590	9588	9587	9586	9585
.46	9584	9583	9582	9581	9579	9578	9576	9575	9574	9573
.47	9572	9571	9570	9569	9567	9566	9564	9563	9562	9561
.48	9560	9559	9558	9556	9555	9553	9552	9550	9549	9548
21.49	9547	9546	9545	9544	9542	9541	9539	9538	9537	9536
In.dec.	0	1	2	3	4	5	6	7	8	9

In.dec.	In. dec. 21.500					In. dec. 21.999				
In.dec.	0	1	2	3	4	5	6	7	8	9
21.50	9535	9534	9533	9532	9530	9529	9527	9526	9525	9524
.51	9523	9522	9521	9520	9518	9517	9515	9514	9513	9512
.52	9511	9510	9509	9508	9506	9505	9503	9502	9501	9500
.53	9499	9498	9497	9496	9494	9493	9491	9490	9489	9488
.54	9487	9486	9485	9484	9482	9481	9479	9478	9477	9476
.55	9475	9474	9473	9472	9470	9469	9467	9466	9465	9464
.56	9463	9462	9461	9460	9458	9457	9455	9454	9453	9452
.57	9451	9450	9449	9448	9446	9445	9443	9442	9441	9440
.58	9439	9438	9437	9436	9434	9433	9431	9430	9429	9428
.59	9427	9426	9425	9424	9422	9421	9419	9418	9417	9416
.60	9415	9414	9413	9412	9410	9409	9407	9406	9405	9404
.61	9403	9402	9401	9399	9398	9396	9395	9393	9392	9391
.62	9390	9389	9388	9387	9385	9384	9382	9381	9380	9379
.63	9378	9377	9376	9375	9373	9372	9370	9369	9368	9367
.64	9366	9365	9364	9363	9361	9360	9358	9357	9356	9355
.65	9354	9353	9352	9351	9349	9348	9346	9345	9344	9343
.66	9342	9341	9340	9338	9337	9336	9334	9333	9332	9331
.67	9330	9329	9328	9327	9325	9324	9322	9321	9320	9319
.68	9318	9317	9316	9315	9313	9312	9310	9309	9308	9307
.69	9306	9305	9304	9303	9301	9300	9298	9297	9296	9295
.70	9294	9293	9292	9291	9289	9288	9286	9285	9284	9283
.71	9282	9281	9280	9279	9277	9276	9274	9273	9272	9271
.72	9270	9269	9268	9267	9265	9264	9262	9261	9260	9259
.73	9258	9257	9256	9255	9253	9252	9250	9249	9248	9247
74	9246	9245	9244	9243	9241	9240	9238	9237	9236	9235
21.75	9234	9233	9232	9231	9229	9228	9226	9225	9224	9223
76	9222	9221	9220	9219	9217	9216	9214	9213	9212	9211
.77	9210	9209	9208	9207	9205	9204	9202	9201	9200	9199
.78	9198	9197	9196	9195	9193	9192	9190	9189	9188	9187
.79	9186	9185	9184	9183	9181	9180	9178	9177	9176	9175
.80	9174	9173	9172	9171	9169	9168	9166	9165	9164	9163
.81	9162	9161	9160	9159	9158	9156	9155	9154	9153	9152
.82	9151	9150	9149	9148	9146	9145	9143	9142	9141	9140
.83	9139	9138	9137	9136	9134	9133	9131	9130	9129	9128
.84	9127	9126	9125	9124	9122	9121	9119	9118	9117	9116
.85	9115	9114	9113	9112	9110	9109	9107	9106	9105	9104
.86	9103	9102	9101	9100	9098	9097	9095	9094	9093	9092
.87	9091	9090	9089	9088	9086	9085	9083	9082	9081	9080
.88	9079	9078	9077	9076	9074	9073	9071	9070	9069	9068
.89	9067	9066	9065	9064	9062	9061	9059	9058	9057	9056
.90	9055	9054	9053	9052	9050	9049	9047	9046	9045	9044
.91	9043	9042	9041	9040	9038	9037	9035	9034	9033	9032
.92	9031	9030	9029	9028	9026	9025	9023	9022	9021	9020
.93	9019	9018	9017	9016	9014	9013	9011	9010	9009	9008
.94	9007	9006	9005	9004	9003	9001	9000	8999	8998	8997
.95	8996	8995	8994	8993	8991	8990	8988	8987	8986	8985
.96	8984	8983	8982	8981	8979	8978	8976	8975	8974	8973
.97	8972	8971	8970	8969	8967	8966	8964	8963	8962	8961
.98	8960	8959	8958	8957	8955	8954	8952	8951	8950	8949
21.99	8948	8947	8946	8945	8943	8942	8940	8939	8938	8937
In dec.	0	1	2	3	4	5	6	7	8	9

In. dec. 22 000					In. dec. 22.499				
ln.dec.	**0**	**1**	**2**	**3**	**4**	**5**	**6**	**7**	**8**	**9**

ln.dec.	0	1	2	3	4	5	6	7	8	9
22.00	8936	8935	8934	8933	8931	8930	8929	8927	8926	8925
.01	8924	8923	8922	8921	8919	8918	8917	8915	8914	8913
.02	8912	8911	8910	8909	8908	8906	8905	8904	8903	8902
.03	8901	8900	8899	8898	8896	8895	8894	8892	8891	8890
.04	8889	8888	8887	8886	8884	8883	8882	8880	8879	8878
.05	8877	8876	8875	8874	8872	8871	8870	8868	8867	8866
.06	8865	8864	8863	8862	8861	8859	8858	8857	8856	8855
.07	8854	8853	8852	8851	8849	8848	8847	8845	8844	8843
.08	8842	8841	8840	8839	8837	8836	8835	8833	8832	8831
.09	8830	8829	8828	8827	8825	8824	8823	8821	8820	8819
.10	8818	8817	8816	8815	8813	·8812	8811	8809	8808	8807
.11	8806	8805	8804	8803	8801	8800	8799	8797	8796	8795
.12	8794	8793	8792	8791	8790	8788	8787	8786	8785	8784
.13	8783	8782	8781	8780	8778	8777	8776	8774	8773	8772
.14	8771	8770	8769	8768	8766	8765	8764	8762	8761	8760
.15	8759	8758	8757	8756	8754	8753	8752	8750	8749	8748
.16	8747	8746	8745	8744	8742	8741	8740	8738	8737	8736
.17	8735	8734	8733	8732	8731	8729	8728	8727	8726	8725
.18	8724	8723	8722	8721	8719	8718	8717	8715	8714	8713
.19	8712	8711	8710	8709	8708	8706	8705	8704	8703	8702
.20	8701	8700	8699	8698	8696	8695	8694	8692	8691	8690
.21	8689	8688	8687	8686	8684	8683	8682	8680	8679	8678
.22	8677	8676	8675	8674	8672	8671	8670	8668	8667	8666
.23	8665	8664	8663	8662	8661	8659	8658	8657	8656	8655
.24	8654	8653	8652	8651	8649	8648	8647	8645	8644	8643
22.25	8642	8641	8640	8639	8637	8636	8635	8633	8632	8631
.26	8630	8629	8628	8627	8625	8624	8623	8621	8620	8619
.27	8618	8617	8616	8615	8613	8612	8611	8609	8608	8607
.28	8606	8605	8604	8603	8602	8600	8599	8598	8597	8596
.29	8595	8594	8593	8592	8591	8589	8588	8587	8596	8585
.30	8584	8583	8582	8581	8579	8578	8577	8575	8574	8573
.31	8572	8571	8570	8569	8567	8566	8565	8563	8562	8561
.32	8560	8559	8558	8557	8555	8554	8553	8551	8550	8549
.33	8548	8547	8546	8545	8544	8542	8541	8540	8539	8538
.34	8537	8536	8535	8534	8532	8531	8530	8528	8527	8526
.35	8525	8524	8523	8522	8520	8519	8518	8516	8515	8514
.36	8513	8512	8511	8510	8508	8507	8505	8504	8503	8502
.37	8501	8500	8499	8498	8497	8495	8494	8493	8492	8491
.38	8490	8489	8488	8487	8486	8484	8483	8482	8481	8480
.39	8479	8478	8477	8476	8474	8473	8471	8470	8469	8468
.40	8467	8466	8465	8464	8462	8461	8459	8458	8457	8456
.41	8455	8454	8453	8452	8450	8449	8447	8446	8445	8444
.42	8443	8442	8441	8440	8439	8437	8436	8435	8434	8433
.43	8432	8431	8430	8429	8427	8426	8424	8423	8422	8421
.44	8420	8419	8418	8417	8415	9414	8412	8411	841c	8409
.45	8408	8407	8406	8405	8404	8402	8401	8400	8399	8398
.46	8397	8396	8395	8394	8393	8391	8390	8388	8388	8387
.47	8386	8385	8384	8383	8381	8380	8378	8377	8376	8375
.48	8374	8373	8372	8371	8369	8368	8366	8365	8364	8363
22.49	8362	8361	8360	8359	8358	8356	8355	8354	8353	8352
ln.dec.	0	1	2	3	4	5	6	7	8	9

	In. dec. 22.500					In. dec. 22 999				
In.dec.	0	1	2	3	4	5	6	7	8	9
22.50	8351	8350	8349	8348	8346	8345	8344	8342	8341	8340
.51	8339	8338	8337	8336	8334	8333	8332	8330	8329	8328
.52	8327	8326	8325	8324	8323	8321	8320	8319	8318	8317
.53	8316	8315	8314	8313	8312	8310	8309	8308	8307	8306
.54	8305	8204	8203	8302	8300	8299	8298	8296	8295	8294
.55	8293	8292	8291	8290	8288	8287	8286	8284	8283	8282
.56	8281	8280	8279	8278	8277	8275	8274	8273	8272	8271
.57	8270	8269	8268	8267	8265	8264	8263	8261	8260	8259
.58	8258	8257	8256	8255	8253	8252	8251	82.9	8248	8247
.59	8246	8245	8244	8243	8242	8240	8239	8238	8237	8236
.60	8235	8234	8233	8232	8231	8229	8228	8227	8226	8225
.61	8224	8223	8222	8221	8219	8218	8217	8215	8214	8213
.62	8212	8211	8210	8209	8208	8206	8205	8204	8203	8202
.63	8201	8200	8199	8198	8196	8195	8194	8192	8191	8190
.64	8189	8188	8187	8186	8184	8183	8182	8180	8179	8178
.65	8177	8176	8175	8174	8173	8171	8170	8169	8168	8167
.66	8166	8165	8164	8163	8162	8160	8159	8158	8157	8156
.67	8155	8154	8153	8152	8150	8149	8148	8146	8145	8144
.68	8143	8142	8141	8140	8139	8137	8136	8135	8134	8133
.69	8132	8131	8130	8129	8127	8126	8125	8123	8122	8121
.70	8120	8119	8118	8117	8115	8114	8113	8111	8110	8109
.71	8108	8107	8106	8105	8104	8102	8101	8100	8099	8098
.72	8097	8096	8095	8094	8093	8091	8090	8089	8088	8087
.73	8086	8085	8084	8083	8081	8080	8079	8077	8076	8075
.74	8074	8073	8072	8071	8070	8068	8067	8066	8065	8064
22.75	8063	8062	8061	8060	8058	8057	8056	8054	8053	8052
.76	8051	8050	8049	8048	8047	8045	8044	8043	8042	8041
.77	8040	8039	8038	8037	8036	8034	8033	8032	8031	8030
.78	8029	8028	8027	8026	8024	8023	8022	8020	8019	8018
.79	8017	8016	8015	8014	8013	8011	8010	8009	8008	8007
.80	8006	8005	8004	8003	8001	8000	7999	7997	7996	7995
.81	7994	7993	7992	7991	7989	7988	7987	7985	7984	7983
.82	7982	7981	7980	7979	7978	7976	7975	7974	7973	7972
.83	7971	7970	7969	7968	7967	7965	7964	7963	7962	7961
.84	7960	7959	7958	7957	7955	7954	7953	7951	7950	7949
.85	7948	7947	7946	7945	7944	7942	7941	7940	7939	7938
.86	7937	7936	7935	7934	7932	7931	7930	7928	7927	7926
.87	7925	7924	7923	7922	7921	7919	7918	7917	7916	7915
.88	7914	7913	7912	7911	7910	7908	7907	7906	7905	7904
.89	7903	7902	7901	7900	7898	7897	7896	7894	7893	7892
.90	7891	7890	7889	7888	7887	7885	7884	7883	7882	7881
.91	7880	7879	7878	7877	7876	7874	7873	7872	7871	7870
.92	7869	7868	7867	7866	7865	7863	7862	7861	7860	7859
.93	7858	7857	7856	7855	7853	7852	7851	7849	7848	7847
.94	7846	7845	7844	7843	7842	7840	7839	7838	7837	7836
.95	7835	7834	7833	7832	7830	7829	7828	7826	7825	7824
.96	7823	7822	7821	7820	7819	7817	7816	7815	7814	7813
.97	7812	7811	7810	7809	7808	7806	7805	7804	7803	7802
.98	7801	7800	7799	7798	3796	7795	7794	7792	7791	7790
22.99	7789	7788	7787	7786	7785	7783	7782	7781	7780	7779
In.dec.	0	1	2	3	4	5	6	7	8	9

	In. dec. 23.000					In. dec. 23.499				
In.dec.	0	1	2	3	4	5	6	7	8	9
23.00	7778	7777	7776	7775	7773	7772	7771	7769	7768	7767
.01	7766	7765	7764	7763	7762	7760	7759	7758	7757	7756
.02	7755	7754	7753	7752	7751	7749	7748	7747	7746	7745
.03	7744	7743	7742	7741	7740	7738	7737	7736	7735	7734
.04	7733	7732	7731	7730	7728	7727	7726	7724	7723	7722
.05	7721	7720	7719	7718	7717	7715	7714	7713	7712	7711
.06	7710	7709	7708	7707	7706	7704	7703	7702	7701	7700
.07	7699	7698	7697	7696	7694	7693	7692	7690	7689	7688
.08	7687	7686	7685	7684	7683	7681	7680	7679	7678	7677
.09	7676	7675	7674	7673	7672	7670	7669	7668	7667	7666
.10	7665	7664	7663	7662	7661	7659	7658	7657	7656	7655
.11	7654	7653	7652	7651	7649	7648	7647	7645	7644	7643
.12	7642	7641	7640	7639	7638	7636	7635	7634	7633	7632
.13	7631	7630	7629	7628	7627	7625	7624	7623	7622	7621
.14	7620	7619	7618	7617	7616	7614	7613	7612	7611	7610
.15	7609	7608	7607	7606	7604	7603	7602	7600	7599	7598
.16	7597	7596	7595	7594	7593	7591	7590	7589	7588	7587
.17	7586	7585	7584	7583	7582	7580	7579	7578	7577	7576
.18	7575	7574	7573	7572	7571	7569	7568	7567	7566	7565
.19	7564	7563	7562	7561	7559	7558	7557	7555	7554	7553
.20	7552	7551	7550	7549	7548	7546	7545	7544	7543	7542
.21	7541	7540	7539	7538	7537	7535	7534	7533	7532	7531
.22	7530	7529	7528	7527	7526	7524	7523	7522	7521	7520
.23	7519	7518	7517	7516	7514	7513	7512	7510	7509	7508
.24	7507	7506	7505	7504	7503	7501	7500	7499	7498	7497
23.25	7496	7495	7494	7493	7492	7490	7489	7488	7487	7486
.26	7485	7484	7483	7482	7481	7479	7478	7477	7476	7475
.27	7474	7473	7472	7471	7470	7468	7467	7466	7465	7464
.28	7463	7462	7461	7460	7458	7457	7456	7454	7453	7452
.29	7451	7450	7449	7448	7447	7445	7444	7443	7442	7441
.30	7440	7439	7438	7437	7436	7434	7433	7432	7431	7430
.31	7429	7428	7427	7426	7425	7423	7422	7421	7420	7419
.32	7418	7417	7416	7415	7414	7412	7411	7410	7409	7408
.33	7407	7406	7405	7404	7403	7401	7400	7399	7398	7397
.34	7396	7395	7394	7393	7391	7390	7389	7387	7386	7385
.35	7384	7383	7382	7381	7380	7378	7377	7376	7375	7374
.36	7373	7372	7371	7370	7369	7367	7366	7365	7364	7363
.37	7362	7361	7360	7359	7358	7356	7355	7354	7353	7352
.38	7351	7350	7349	7348	7347	7345	7344	7343	7342	7341
.39	7340	7339	7338	7337	7335	7334	7333	7331	7330	7329
.40	7328	7327	7326	7325	7324	7322	7321	7320	7319	7318
.41	7317	7316	7315	7314	7313	7311	7310	7309	7308	7307
.42	7306	7305	7304	7303	7302	7300	7299	7298	7297	7296
.43	7295	7294	7293	7292	7291	7289	7288	7287	7286	7285
.44	7284	7283	7282	7281	7280	7278	7277	7276	7275	7274
.45	7273	7272	7271	7270	7269	7267	7266	7265	7264	7263
.46	7262	7261	7260	7259	7258	7256	7255	7254	7253	7252
.47	7251	7250	7249	7248	7247	7245	7244	7243	7242	7241
.48	7240	7239	7238	7237	7236	7234	7233	7232	7231	7230
23.49	7229	7228	7227	7226	7224	7223	7222	7220	7219	7218
In.dec.	0	1	2	3	4	5	6	7	8	9

D

	In. dec. 23.500					In. dec. 23.999				
In.dec.	0	1	2	3	4	5	6	7	8	9
23.50	7217	7216	7215	7214	7213	7211	7210	7209	7208	7207
.51	7206	7205	7204	7203	7202	7200	7199	7198	7197	7196
.52	7195	7194	7193	7192	7191	7189	7188	7187	7186	7185
.53	7184	7183	7182	7181	7180	7178	7177	7176	7175	7174
.54	7173	7172	7171	7170	7169	7167	7166	7165	7164	7163
.55	7162	7161	7160	7159	7158	7156	7155	7154	7153	7152
.56	7151	7150	7159	7148	7147	7145	7144	7143	7142	7141
.57	7140	7139	7138	7137	7136	7134	7133	7132	7131	7130
.58	7129	7128	7127	7126	7125	7123	7122	7121	7120	7119
.59	7118	7117	7116	7115	7114	7112	7111	7110	7109	7108
.60	7107	7106	7105	7104	7103	7101	7100	7099	7098	7097
.61	7096	7095	7094	7093	7092	7090	7089	7088	7087	7086
.62	7085	7084	7083	7082	7081	7079	7078	7077	7076	7075
.63	7074	7073	7072	7071	7070	7068	7067	7066	7065	7064
.64	7063	7062	7061	7060	7059	7057	7056	7055	7054	7053
.65	7052	7051	7050	7049	7048	7046	7045	7044	7043	7042
.66	7041	7040	7039	7038	7037	7035	7034	7033	7032	7031
.67	7030	7029	7028	7027	7026	7024	7023	7022	7021	7020
.68	7019	7018	7017	7016	7015	7013	7012	7011	7010	7009
.69	7008	7007	7006	7005	7004	7002	7001	7000	6999	6998
.70	6997	6996	6995	6994	6993	6991	6990	6989	6988	6987
.71	6986	6985	6984	6983	6982	6980	6979	6978	6977	6976
.72	6975	6974	6973	6972	6971	6969	6968	6967	6966	6965
.73	6964	6963	6962	6961	6960	6958	6957	6956	6955	6954
.74	6953	6952	6951	6950	6949	6947	6946	6945	6944	6943
23.75	6942	6941	6940	6939	6938	6936	6935	6934	6933	6932
.76	6931	6930	6929	6928	6927	6925	6924	6923	6922	6921
.77	6920	6919	6918	6917	6916	6914	6913	6912	6911	6910
.78	6909	6908	6907	6906	6905	6903	6902	6901	6900	6899
79	6898	6897	6896	6895	6894	6892	6891	6890	6889	6888
.80	6887	6886	6885	6884	6883	6881	6880	6879	6878	6877
.81	6876	6875	6874	6873	6872	6870	6869	6868	6867	6866
.82	6865	6864	6863	6862	6861	6859	6858	6857	6856	6855
.83	6854	6853	6852	6851	6850	6848	6847	6846	6845	6844
.84	6843	6842	6841	6840	6839	6837	6836	6835	6834	6833
.85	6832	6831	6830	6829	6828	6826	6825	6824	6823	6822
.86	6821	6820	6819	6818	6817	6816	6815	6814	6813	6812
.87	6811	6810	6809	6808	6808	6805	6804	6803	6802	6801
.88	6800	6799	6798	6797	6795	6794	6793	6791	6790	6789
.89	6788	6787	6786	6785	6784	6782	6781	6780	6779	6778
.90	6777	6776	6775	6774	6773	6772	6771	6770	6769	6768
.91	6767	6766	6765	6764	6763	6761	6760	6759	6758	6757
.92	6756	6755	6754	6753	6752	6750	6749	6748	6747	6746
.93	6745	6744	6743	6742	6741	6739	6738	6737	6736	6735
.94	6734	6733	6732	6731	6730	6728	6727	6726	6725	6724
.95	6723	6722	6721	6720	6719	6717	6716	6715	6714	6713
.96	6712	6711	6710	6709	6708	6706	6705	6704	6703	6702
.97	6701	6700	6699	6698	6697	6695	6694	6693	6692	6691
.98	6690	6689	6688	6687	6686	6685	6684	6683	6682	6681
23.99	6680	6679	6678	6677	6676	6674	6673	6672	6671	6670
In.dec.	0	1	2	3	4	5	6	7	8	9

	In. dec. 24.000					In. dec. 24.499				
In.dec.	0	1	2	3	4	5	6	7	8	9
24.00	6669	6668	6667	6666	6665	6663	6662	6661	6660	6659
.01	6658	6657	6656	6655	6654	6652	6651	6650	6649	6648
.02	6647	6646	6645	6644	6643	6642	6641	6640	6639	6638
.03	6637	6636	6635	6634	6633	6631	6630	6629	6628	6627
.04	6626	6625	6624	6623	6622	6620	6619	6618	6617	6616
.05	6615	6614	6613	6612	6611	6609	6608	6607	6606	6605
.06	6604	6603	6602	6601	6600	6598	6597	6596	6595	6594
.07	6593	6592	6591	6590	6589	6587	6586	6585	6584	6583
.08	6582	6581	6580	6579	6578	6576	6575	6574	6573	6572
.09	6571	6570	6569	6568	6567	6565	6564	6563	6562	6561
.10	6560	6559	6558	6557	6556	6555	6554	6553	6552	6551
.11	6550	6549	6548	6547	6546	6544	6543	6542	6541	6540
.12	6539	6538	6537	6536	6535	6533	6532	6531	6530	6529
.13	6528	6527	6526	6525	6524	6522	6521	6520	6519	6518
.14	6517	6516	6515	6514	6513	6511	6510	6509	6508	6507
.15	6506	6505	6504	6503	6502	6501	6500	6499	6498	6497
.16	6496	6495	6494	6493	6492	6490	6489	6488	6487	6486
.17	6485	6484	6483	6482	6481	6479	6478	6477	6476	6475
.18	6474	6473	6472	6471	6470	6468	6467	6466	6465	6464
.19	6463	6462	6461	6460	6459	6457	6456	6455	6454	6453
.20	6452	6451	6450	6449	6448	6447	6446	6445	6444	6443
.21	6442	6441	6440	6439	6438	6436	6435	6434	6433	6432
.22	6431	6430	6429	6428	6427	6426	6425	6424	6423	6422
.23	6421	6420	6419	6418	6417	6415	6414	6413	6412	6411
.24	6410	6409	6408	6407	6406	6404	6403	6402	6401	6400
24.25	6399	6398	6397	6396	6395	6393	6392	6391	6390	6389
.26	6388	6387	6386	6385	6384	6382	6381	6380	6379	6378
.27	6377	6376	6375	6374	6373	6372	6371	6370	6369	6368
.28	6367	6366	6365	6364	6363	6361	6360	6359	6358	6357
.29	6356	6355	6354	6353	6352	6350	6349	6348	6347	6346
.30	6345	6344	6343	6342	6341	6340	6339	6338	6337	6336
.31	6335	6334	6333	6332	6331	6329	6328	6327	6326	6325
.32	6324	6323	6322	6321	6320	6318	6317	6316	6315	6314
.33	6313	6312	6311	6310	6309	6307	6306	6305	6304	6303
.34	6302	6301	6300	6299	6298	6297	6296	6295	6294	6293
.35	6292	6291	6290	6289	6288	6286	6285	6284	6283	6282
.36	6281	6280	6279	6278	6277	6275	6274	6273	6272	6271
.37	6270	6269	6268	6267	6266	6265	6264	6263	6262	6261
.38	6260	6259	6258	6257	6256	6254	6253	6252	6251	6250
.39	6249	6248	6247	6246	6245	6243	6242	6241	6240	6239
.40	6238	6237	6236	6235	6234	6232	6231	6230	6229	6228
.41	6227	6226	6225	6224	6223	6222	6221	6220	6219	6218
.42	6217	6216	6215	6214	6213	6211	6210	6209	6208	6207
.43	6206	6205	6204	6203	6202	6201	6200	6199	6198	6197
.44	6196	6195	6194	6193	6192	6190	6189	6188	6187	6186
.45	6185	6184	6183	6182	6181	6179	6178	6177	6176	6175
.46	6174	6173	6172	6171	6170	6169	6168	6167	6166	6165
.47	6164	6163	6162	6161	6160	6158	6157	6156	6155	6154
.48	6153	6152	6151	6150	6149	6147	6146	6145	6144	6143
24.49	6142	6141	6140	6139	6138	6136	6135	6134	6133	6132
In.dec.	0	1	2	3	4	5	6	7	8	9

In.dec.	In. dec. 24.500					In. dec. 24.999				
	0	1	2	3	4	5	6	7	8	9
24.50	6131	6130	6129	6128	6127	6126	6125	6124	6123	6122
.51	6121	6120	6119	6118	6117	6115	6114	6113	6112	6111
.52	6110	6109	6108	6107	6106	6105	6104	6103	6102	6101
.53	6100	6099	6098	6097	6096	6094	6093	6092	6091	6090
:54	6089	6088	6087	6086	6085	6084	6083	6082	6081	6080
.55	6079	6078	6077	6076	6075	6073	6072	6071	6070	6069
.56	6068	6067	6066	6065	6064	6062	6061	6060	6059	6058
.57	6057	6056	6055	6054	6053	6052	6051	6050	6049	6048
.58	6047	6046	6045	6044	6043	6041	6040	6039	6038	6037
.59	6036	6035	6034	6033	6032	6030	6029	6028	6027	6026
.60	6025	6024	6023	6022	6021	6020	6019	6018	6017	6016
.61	6015	6014	6013	6012	6011	6009	6008	6007	6006	6005
.62	6004	6003	6002	6001	6000	5999	5998	5997	5996	5995
.63	5994	5993	5992	5991	5990	5988	5987	5986	5985	5984
.64	5983	5982	5981	5980	5979	5977	5976	5975	5974	5973
.65	5972	5971	5970	5969	5968	5967	5966	5965	5964	5963
.66	5962	5961	5960	5959	5958	5956	5955	5954	5953	5952
.67	5951	5950	5949	5948	5947	5946	5945	5944	5943	5942
.68	5941	5940	5939	5938	5937	5935	5934	5933	5932	5931
.69	5930	5929	5928	5927	5926	5925	5924	5923	5922	5921
.70	5920	5919	5918	5917	5916	5915	5914	5913	5912	5911
.71	5910	5909	5908	5907	5906	5904	5903	5902	5901	5900
72	5899	5898	5897	5896	5895	5893	5892	5891	5890	5889
.73	5888	5887	5886	5885	5884	5883	5882	5881	5880	5879
.74	5878	5877	5876	5875	5874	5872	5871	5870	5869	5868
24.75	5867	5866	5865	5864	5863	5862	5861	5860	5859	5858
76	5857	5856	5855	5854	5853	5851	5850	5849	5848	5847
.77	5846	5845	5844	5843	5842	5841	5840	5839	5838	5837
.78	5836	5835	5834	5833	5832	5830	5829	5828	5827	5826
.79	5825	5824	5823	5822	5821	5820	5819	5818	5817	5816
.80	5815	5814	5813	5812	5811	5809	5808	5807	5806	5805
.81	5804	5803	5802	5801	5800	5799	5798	5797	5796	5795
.82	5794	5793	5792	5791	5790	5788	5787	5786	5785	5784
.83	5783	5782	5781	5780	5779	5778	5777	5776	5775	5774
.84	5773	5772	5771	5770	5769	5767	5766	5765	5764	5763
.85	5762	5761	5760	5759	5758	5757	5756	5755	5754	5753
.86	5752	5751	5750	5749	5748	5746	5745	5744	5743	5742
.87	5741	5740	5739	5738	5737	5736	5735	5734	5733	5732
.88	5731	5730	5729	5728	5727	5725	5724	5723	5722	5721
.89	5720	5719	5718	5717	5716	5715	5714	5713	5712	5711
.90	5710	5709	5708	5707	5706	5704	5703	5702	5701	5700
.91	5699	5698	5697	5696	5695	5694	5693	5692	5691	5690
.92	5689	5688	5687	5686	5685	5683	5682	5681	5680	5679
.93	5678	5677	5676	5675	5674	5673	5672	5671	5670	5669
.94	5668	5667	5666	5665	5664	5662	5661	5660	5659	5658
.95	5657	5656	5655	5654	5653	5652	5651	5650	5649	5648
.96	5647	5646	5645	5644	5643	5641	5640	5639	5638	5637
.97	5636	5635	5634	5633	5632	5631	5630	5629	5628	5627
.98	5626	5625	5624	5623	5622	5620	5619	5618	5617	5616
24.99	5615	5614	5613	5612	5611	5610	5609	5608	5607	5606
In.dec.	0	1	2	3	4	5	6	7	8	9

n.dec.	In. dec. 25.000					In. dec. 25.499				
	0	1	2	3	4	5	6	7	8	9
25.00	5605	5604	5603	5602	5601	5600	5599	5598	5597	5596
.01	5595	5594	5593	5592	5591	5589	5588	5587	5586	5585
.02	5584	5583	5582	5581	5580	5579	5578	5577	5576	5575
.03	5574	5573	5572	5571	5570	5569	5568	5567	5566	5565
.04	5564	5563	5562	5561	5560	5558	5557	5556	5555	5554
.05	5553	5552	5551	5550	5549	5548	5547	5546	5545	5544
.06	5543	5542	5541	5540	5539	5538	5537	5536	5535	5534
.07	5533	5532	5531	5530	5529	5527	5526	5525	5524	5523
.08	5522	5521	5520	5519	5518	5517	5516	5515	5514	5513
.09	5512	5511	5510	5509	5508	5506	5505	5504	5503	5502
.10	5501	5500	5499	5498	5497	5496	5495	5494	5493	5492
.11	5491	5490	5489	5488	5487	5485	5484	5483	5482	5481
.12	5480	5479	5478	5477	5476	5475	5474	5473	5472	5471
.13	5470	5469	5468	5467	5466	5464	5463	5462	5461	5460
.14	5459	5458	5457	5456	5455	5454	5453	5452	5451	5450
.15	5449	5448	5447	5446	5445	5444	5443	5442	5441	5440
.16	5439	5438	5437	5436	5435	5434	5433	5432	5431	5430
.17	5429	5428	5427	5426	5425	5423	5422	5421	5420	5419
18	5418	5417	5416	5415	5414	5413	5412	5411	5410	5409
.19	5408	5407	5406	5405	5404	5403	5402	5401	5400	5399
.20	5398	5397	5396	5395	5394	5392	5391	5390	5389	5388
.21	5387	5386	5385	5384	5383	5382	5381	5380	5379	5378
.22	5377	5376	5375	5374	5373	5371	5370	5369	5368	5367
.23	5366	5365	5364	5363	5362	5361	5360	5359	5358	5357
.24	5356	5355	5354	5353	5352	5351	5350	5349	5348	5347
25.25	5346	5345	5344	5343	5342	5341	5340	5339	5338	5337
.26	5336	5335	5334	5333	5332	5330	5329	5328	5327	5326
.27	5325	5324	5323	5322	5321	5320	5319	5318	5317	5316
.28	5315	5314	5313	5312	5311	5310	5309	5308	5307	5306
.29	5305	5304	5303	5302	5301	5299	5298	5297	5296	5295
.30	5294	5293	5292	5291	5290	5289	5288	5287	5286	5285
.31	5284	5283	5282	5281	5280	5279	5278	5277	5276	5275
.32	5274	5273	5272	5271	5270	5268	5267	5266	5265	5264
.33	5263	5262	5261	5260	5259	5258	5257	5256	5255	5254
.34	5253	5252	5251	5250	5249	5248	5247	5246	5245	5244
.35	5243	5242	5241	5240	5239	5238	5237	5236	5235	5234
.36	5233	5232	5231	5230	5229	5227	5226	5225	5224	5223
.37	5222	5221	5220	5219	5218	5217	5216	5215	5214	5213
.38	5212	5211	5210	5209	5208	5207	5206	5205	5204	5203
..39	5202	5201	5200	5199	5198	5197	5196	5195	5194	5193
.40	5192	5191	5190	5189	5188	5187	5186	5185	5184	5183
.41	5182	5181	5180	5179	5178	5176	5175	5174	5173	5172
.42	5171	5170	5169	5168	5167	5166	5165	5164	5163	5162
.43	5161	5160	5159	5158	5157	5155	5154	5153	5152	5151
.44	5150	5149	5148	5147	5146	5145	5144	5143	5142	5141
.45	5140	5139	5138	5137	5136	5135	5134	5133	5132	5131
.46	5130	5129	5128	5127	5126	5125	5124	5123	5122	5121
.47	5120	5119	5118	5117	5116	5115	5114	5113	5112	5111
.48	5110	5109	5108	5107	5106	5104	5103	5102	5101	5100
25.49	5099	5098	5097	5096	5095	5094	5093	5092	5091	5090
n.dec.	0	1	2	3	4	5	6	7	8	9

	In. dec. 25.500					In. dec. 25.999				
In.dec.	0	1	2	3	4	5	6	7	8	9
25.50	5089	5088	5087	5086	5085	5084	5083	5082	5081	5080
.51	5079	5078	5077	5076	5075	5074	5073	5072	5071	5070
.52	5069	5068	5067	5066	5065	5064	5063	5062	5061	5060
.53	5059	5058	5057	5056	5055	5053	5052	5051	5050	5049
.54	5048	5047	5046	5045	5044	5043	5042	5041	5040	5039
.55	5038	5037	5036	5035	5034	5033	5032	5031	5030	5029
.56	5028	5027	5026	5025	5024	5023	5022	5021	5020	5019
.57	5018	5017	5016	5015	5014	5013	5012	5011	5010	5009
.58	5008	5007	5006	5005	5004	5002	5001	5000	4999	4998
.59	4997	4996	4995	4994	4993	4992	4991	4990	4989	4988
.60	4987	4986	4985	4984	4983	4982	4981	4980	4979	4978
.61	4977	4976	4975	4974	4973	4972	4971	4970	4969	4968
.62	4967	4966	4965	4964	4963	4962	4961	4960	4959	4958
.63	4957	4956	4955	4954	4953	4951	4950	4949	4948	4947
.64	4946	4945	4944	4943	4942	4941	4940	4939	4938	4937
.65	4936	4935	4934	4933	4932	4931	4930	4929	4928	4927
.66	4926	4925	4924	4923	4922	4921	4920	4919	4918	4917
.67	4916	4915	4914	4913	4912	4911	4910	4909	4908	4907
.68	4906	4905	4904	4903	4902	4901	4900	4899	4898	4897
.69	4896	4895	4894	4893	4892	4891	4890	4889	4888	4887
.70	4886	4885	4884	4883	4882	4881	4880	4879	4878	4877
.71	4876	4875	4874	4873	4872	4870	4869	4868	4867	4866
.72	4865	4864	4863	4862	4861	4860	4859	4858	4857	4856
73	4855	4854	4853	4852	4851	4850	4849	4848	4847	4846
.74	4845	4844	4843	4842	4841	4840	4839	4838	4837	4836
25.75	4835	4834	4833	4832	4831	4830	4829	4828	4827	4826
.76	4825	4824	4823	4822	4821	4820	4819	4818	4817	4816
.77	4815	4814	4813	4812	4811	4810	4809	4808	4807	4806
.78	4805	4804	4803	4802	4801	4800	4799	4798	4797	4796
79	4795	4794	4793	4792	4791	4789	4788	4787	4786	4785
.80	4784	4783	4782	4781	4780	4779	4778	4777	4776	4775
.81	4774	4773	4772	4771	4770	4769	4768	4767	4766	4765
.82	4764	4763	4762	4761	4760	4759	4758	4757	4756	4755
.83	4754	4753	4752	4751	4750	4749	4748	4747	4746	4745
.84	4744	4743	4742	4741	4740	4739	4738	4737	4736	4735
.85	4734	4733	4732	4731	4730	4729	4728	4727	4726	4725
.86	4724	4723	4722	4721	4720	4719	4718	4717	4716	4715
.87	4714	4713	4712	4711	4710	4709	4708	4707	4706	4705
.88	4704	4703	4702	4701	4700	4699	4698	4697	4696	4695
.89	4694	4693	4692	4691	4690	4689	4688	4687	4686	4685
.90	4684	4683	4682	4681	4680	4679	4678	4677	4676	4675
.91	4674	4673	4672	4671	4670	4669	4668	4667	4666	4665
.92	4664	4663	4662	4661	4660	4659	4658	4657	4656	4655
.93	4654	4653	4652	4651	4650	4648	4647	4646	4645	4644
.94	4643	4642	4641	4640	4639	4638	4637	4636	4635	4634
.95	4633	4632	4631	4630	4629	4628	4627	4626	4625	4624
.96	4623	4622	4621	4620	4619	4618	4617	4616	4615	4614
.97	4613	4612	4611	4610	4609	4608	4607	4606	4605	4604
.98	4603	4602	4601	4600	4599	4598	4597	4596	4595	4594
25.99	4593	4592	4591	4590	4589	4588	4587	4586	4585	4584
In.dec.	0	1	2	3	4	5	6	7	8	9

In. dec. 26.000						In. dec. 26.499			

In dec	0	1	2	3	4	5	6	7	8	9
26.00	4583	4582	4581	4580	4579	4578	4577	4576	4575	4574
.01	4573	4572	4571	4570	4569	4568	4567	4566	4565	4564
.02	4563	4562	4561	4560	4559	4558	4557	4556	4555	4554
.03	4553	4552	4551	4550	4549	4548	4547	4546	4545	4544
.04	4543	4542	4541	4540	4539	4538	4537	4536	4535	4534
.05	4533	4532	4531	4530	4529	4528	4527	4526	4525	4524
.06	4523	4522	4521	4520	4519	4518	4517	4516	4515	4514
.07	4513	4512	4511	4510	4509	4508	4507	4506	4505	4504
.08	4503	4502	4501	4500	4499	4498	4497	4496	4495	4494
.09	4493	4492	4491	4490	4489	4488	4487	4486	4485	4484
.10	4483	4482	4481	4480	4479	4478	4477	4476	4475	4474
.11	4473	4472	4471	4470	4469	4468	4467	4466	4465	4464
.12	4463	4462	4461	4460	4459	4458	4457	4456	4455	4454
.13	4453	4452	4451	4450	4449	4448	4447	4446	4445	4444
14	4443	4442	4441	4440	4439	4438	4437	4436	4435	4434
.15	4433	4432	4431	4430	4429	4428	4427	4426	4425	4424
.16	4423	4422	4421	4420	4419	4418	4417	4416	4415	4414
.17	4413	4412	4411	4410	4409	4408	4407	4406	4405	4404
18	4403	4402	4401	4400	4399	4398	4397	4396	4395	4394
.19	4393	4392	4391	4390	4389	4388	4387	4386	4385	4384
.20	4383	4382	4381	4380	4379	4378	4377	4376	4375	4374
.21	4373	4372	4371	4370	4369	4368.5	4368	4367	4366	4365
.22	4364	4363	4362	4361	4360	4359	4358	4357	4356	4355
.23	4354	4353	4352	4351	4350	4349	4348	4347	4346	4345
.24	4344	4343	4342	4341	4340	4339	4338	4337	4336	4335
26.25	4334	4333	4332	4331	4330	4329	4328	4327	4326	4325
.26	4324	4323	4322	4321	4320	4319	4318	4317	4316	4315
.27	4314	4313	4312	4311	4310	4309	4308	4307	4306	4305
.28	4304	4303	4302	4301	4300	4299	4298	4297	4296	4295
.29	4294	4293	4292	4291	4290	4289	4288	4287	4286	4285
.30	4284	4283	4282	4281	4280	4279	4278	4277	4276	4275
.31	4274	4273	4272	4271	4270	4269	4268	4267	4266	4265
.32	4264	4263	4262	4261	4260	4259	4258	4257	4256	4255
.33	4254	4253	4252	4251	4250	4249	4248	4247	4246	4245
.34	4244	4243	4242	4241	4240	4239.5	4239	4238	4237	4236
.35	4235	4234	4233	4232	4231	4230	4229	4228	4227	4226
.36	4225	4224	4223	4222	4221	4220	4219	4218	4217	4216
.37	4215	4214	4213	4212	4211	4210	4209	4208	4207	4206
.38	4205	4204	4203	4202	4201	4200	4199	4198	4197	4196
.39	4195	4194	4193	4192	4191	4190	4189	4188	4187	4186
.40	4185	4184	4183	4182	4181	4180	4179	4178	4177	4176
.41	4175	4174	4173	4172	4171	4170.5	4170	4169	4168	4167
.42	4166	4165	4164	4163	4162	4161	4160	4159	4158	4157
.43	4156	4155	4154	4153	4152	4151	4150	4149	4148	4147
.44	4146	4145	4144	4143	4142	4141	4140	4139	4138	4137
.45	4136	4135	4134	4133	4132	4131	4130	4129	4128	4127
.46	4126	4125	4124	4123	4122	4121	4120	4119	4118	4117
.47	4116	4115	4114	4113	4112	4111	4110	4109	4108	4107
.48	4106	4105	4104	4103	4102	4101	4100	4099	4098	4097
26.49	4096	4095	4094	4093	4092	4091.5	4091	4090	4089	4088
In dec.	0	1	2	3	4	5	6	7	8	9

In. dec. 26.500						In. dec. 26.999			
In.dec.	0	1	2	3	4	5	6	7	8	9

In.dec.	0	1	2	3	4	5	6	7	8	9
26.50	4087	4086	4085	4084	4083	4082	4081	4080	4079	4078
.51	4077	4076	4075	4074	4073	4072	4071	4070	4069	4068
.52	4067	4066	4065	4064	4063	4062	4061	4060	4059	4058
.53	4057	4056	4055	4054	4053	4052	4051	4050	4049	4048
.54	4047	4046	4045	4044	4043	4042.5	4042	4041	4040	4039
.55	4038	4037	4036	4035	4034	4033	4032	4031	4030	4029
.56	4028	4027	4026	4025	4024	4023	4022	4021	4020	4019
.57	4018	4017	4016	4015	4014	4013	4012	4011	4010	4009
.58	4008	4007	4006	4005	4004	4003	4002	4001	4000	3999
.59	3998	3997	3996	3995	3994	3993.5	3993	3992	3991	3990
.60	3989	3988	3987	3986	3985	3984	3983	3982	3981	3980
.61	3979	3978	3977	3976	3975	3974	3973	3972	3971	3970
.62	3969	3968	3967	3966	3965	3964	3963	3962	3961	3960
.63	3959	3958	3957	3956	3955	3954.5	3954	3953	3952	3951
.64	3950	3949	3948	3947	3946	3945	3944	3943	3942	3941
.65	3940	3939	3938	3937	3936	3935	3934	3933	3932	3931
.66	3930	3929	3928	3927	3926	3925	3924	3923	3922	3921
.67	3920	3919	3918	3917	3916	3915	3914	3913	3912	3911
.68	3910	3909	3908	3907	3906	3905.5	3905	3904	3903	3902
.69	3901	3900	3899	3898	3897	3896	3895	3894	3893	3892
.70	3891	3890	3889	3888	3887	3886	3885	3884	3883	3882
.71	3881	3880	3879	3878	3877	3876	3875	3874	3873	3872
.72	3871	3870	3869	3868	3867	3866.5	3866	3865	3864	3863
.73	3862	3861	3860	3859	3858	3857	3856	3855	3854	3853
.74	3852	3851	3850	3849	3848	3847	3846	3845	3844	3843
26.75	3842	3841	3840	3839	3838	3837	3836	3835	3834	3833
.76	3832	3831	3830	3829	3828	3827.5	3827	3826	3825	3824
.77	3823	3822	3821	3820	3819	3818	3817	3816	3815	3814
.78	3813	3812	3811	3810	3809	3808	3807	3806	3805	3804
.79	3803	3802	3801	3800	3799	3798.5	3798	3797	3796	3795
.80	3794	3793	3792	3791	3790	3789	3788	3787	3786	3785
.81	3784	3783	3782	3781	3780	3779	3778	3777	3776	3775
.82	3774	3773	3772	3771	3770	3769	3768	3767	3766	3765
.83	3764	3763	3762	3761	3760	3759.5	3759	3758	3757	3756
.84	3755	3754	3753	3752	3751	3750	3749	3748	3747	3746
.85	3745	3744	3743	3742	3741	3740	3739	3738	3737	3736
.86	3735	3734	3733	3732	3731	3730	3729	3728	3727	3726
.87	3725	3724	3723	3722	3721	3720.5	3720	3719	3718	3717
.88	3716	3715	3714	3713	3712	3711	3710	3709	3708	3707
.89	3706	3705	3704	3703	3702	3701.5	3701	3700	3699	3698
.90	3697	3696	3695	3694	3693	3692	3691	3690	3689	3688
.91	3687	3686	3685	3684	3683	3682	3681	3680	3679	3678
.92	3677	3676	3675	3674	3673	3672	3671	3670	3669	3668
.93	3667	3666	3665	3664	3663	3662.5	3662	3661	3660	3659
.94	3658	3657	3656	3655	3654	3653	3652	3651	3650	3649
.95	3648	3647	3646	3645	3644	3643	3642	3641	3640	3639
.96	3638	3637	3636	3635	3634	3633.5	3633	3632	3631	3630
.97	3629	3628	3627	3626	3625	3624	3623	3622	3621	3620
.98	3619	3618	3617	3616	3615	3614.5	3614	3613	3612	3611
26.99	3610	3609	3608	3607	3606	3605	3604	3603	3602	3601
In.dec.	0	1	2	3	4	5	6	7	8	9

	In. dec. 27 000					In. dec. 27.499				
In.dec.	0	1	2	3	4	5	6	7	8	9
27.00	3600	3599	3598	3597	3596	3595	3594	3593	3592	3591
.01	3590	3589	3588	3587	3586	3585	3584	3583	3582	3581
.02	3580	3579	3578	3577	3576	3575.5	3575	3574	3573	3572
.03	3571	3570	3569	3568	3567	3566	3565	3564	3563	3562
.04	3561	3560	3559	3558	3557	3556	3555	3554	3553	3552
.05	3551	3550	3549	3548	3547	3546.5	3546	3545	3544	3543
.06	3542	3541	3540	3539	3538	3537	3536	3535	3534	3533
.07	3532	3531	3530	3529	3528	3527.5	3527	3526	3525	3524
.08	3523	3522	3521	3520	3519	3518	3517	3516	3515	3514
.09	3513	3512	3511	3510	3509	3508	3507	3506	3505	3504
.10	3503	3502	3501	3500	3499	3498.5	3498	3497	3496	3495
.11	3494	3493	3492	3491	3490	3489	3488	3487	3486	3485
.12	3484	3483	3482	3481	3480	3479.5	3479	3478	3477	3476
.13	3475	3474	3473	3472	3471	3470	3469	3468	3467	3466
.14	3465	3464	3463	3462	3461	3460	3459	3458	3457	3456
.15	3455	3454	3453	3452	3451	3450.5	3450	3449	3448	3447
.16	3446	3445	3444	3443	3442	3441	3440	3439	3438	3437
.17	3436	3435	3434	3433	3432	3431.5	3431	3430	3429	3428
.18	3427	3426	3425	3424	3423	3422	3421	3420	3419	3418
.19	3417	3416	3415	3414	3413	3412	3411	3410	3409	3408
.20	3407	3406	3405	3404	3403	3402.5	3402	3401	3400	3399
.21	3398	3397	3396	3395	3394	3393	3392	3391	3390	3389
.22	3388	3387	3386	3385	3384	3383.5	3383	3382	3381	3380
.23	3379	3378	3377	3376	3375	3374	3373	3372	3371	3370
24	3369	3368	3367	3366	3365	3364	3363	3362	3361	3360
27.25	3359	3358	3357	3356	3355	3354.5	3354	3353	3352	3351
.26	3350	3349	3348	3347	3346	3345	3344	3343	3342	3341
.27	3340	3339	3338	3337	3336	3335.5	3335	3334	3333	3332
.28	3331	3330	3329	3328	3327	3326.5	3326	3325	3324	3323
.29	3322	3321	3320	3319	3318	3317	3316	3315	3314	3313
.30	3312	3311	3310	3309	3308	3307	3306	3305	3304	3303
.31	3302	3301	3300	3299	3298	3297.5	3297	3296	3295	3294
.32	3293	3292	3291	3290	3289	3288	3287	3286	3285	3284
.33	3283	3282	3281	3280	3279	3278.5	3278	3277	3276	3275
.34	3274	3273	3272	3271	3270	3269	3268	3267	3266	3265
.35	3264	3263	3262	3261	3260	3259	3258	3257	3256	3255
.36	3254	3253	3252	3251	3250	3249.5	3249	3248	3247	3246
.37	3245	3244	3243	3242	3241	3240.5	3240	3239	3238	3237
.38	3236	3235	3234	3233	3232	3231	3230	3229	3228	3227
.39	3226	3225	3224	3223	3222	3221 5	3221	3220	3219	3218
.40	3217	3216	3215	3214	3213	3212	3211	3210	3209	3208
.41	3207	3206	3205	3204	3203	3202	3201	3200	3199	3198
.42	3197	3196	3195	3194	3193	3192.5	3192	3191	3190	3189
.43	3188	3187	3186	3185	3184	3183.5	3183	3182	3181	3180
.44	3179	3178	3177	3176	3175	3174	3173	3172	3171	3170
.45	3169	3168	3167	3166	3165	3164.5	3164	3163	3162	3161
.46	3160	3159	3158	3157	3156	3155	3154	3153	3152	3151
.47	3150	3149	3148	3147	3146	3145	3144	3143	3142	3141
.48	3140	3139	3138	3137	3136	3135.5	3135	3134	3133	3132
27.49	3131	3130	3129	3128	3127	3126.5	3126	3125	3124	3123
In.dec.	0	1	2	3	4	5	6	7	8	9

E

		In. dec. 27.500					In. dec. 27.999			
In.dec.	0	1	2	3	4	5	6	7	8	9
27.50	3122	3121	3120	3119	3118	3117	3116	3115	3114	3113
.51	3112	3111	3110	3109	3108	3107.5	3107	3106	3105	3104
.52	3103	3102	3101	3100	3099	3098	3097	3096	3095	3094
.53	3093	3092	3091	3090	3089	3088.5	3088	3087	3086	3085
.54	3084	3083	3082	3081	3080	3079	3078	3077	3076	3075
.55	3074	3073	3072	3071	3070	3069.5	3069	3068	3067	3066
.56	3065	3064	3063	3062	3061	3060	3059	3058	3057	3056
.57	3055	3054	3053	3052	3051	3050.5	3050	3049	3048	3047
.58	3046	3045	3044	3043	3042	3041.5	3041	3040	3039	3038
.59	3037	3036	3035	3034	3033	3032	3031	3030	3029	3028
.60	3027	3026	3025	3024	3023	3022	3021	3020	3019	3018
.61	3017	3016	3015	3014	3013	3012.5	3012	3011	3010	3009
62	3008	3007	3006	3005	3004	3003.5	3003	3002	3001	3000
.63	2999	2998	2997	2996	2995	2994	2993	2992	2991	2990
64	2989	2988	2987	2986	2985	2984.5	2984	2983	2982	2981
.65	2980	2979	2978	2977	2976	2975.5	2975	2974	2973	2972
.66	2971	2970	2969	2968	2967	2966	2965	2964	2963	2962
.67	2961	2960	2959	2958	2957	2956	2955	2954	2953	2952
.68	2951	2950	2949	2948	2947	2946.5	2946	2945	2944	2943
.69	2942	2941	2940	2939	2938	2937.5	2937	2936	2935	2934
.70	2933	2932	2931	2930	2929	2928	2927	2926	2925	2924
.71	2923	2922	2921	2920	2919	2918.5	2918	2917	2916	2915
.72	2914	2913	2912	2911	2910	2909.5	2909	2908	2907	2906
.73	2905	2904	2903	2902	2901	2900	2899	2898	2897	2896
.74	2895	2894	2893	2892	2891	2890.5	2890	2889	2888	2887
27.75	2886	2885	2884	2883	2882	2881	2880	2879	2878	2877
76	2876	2875	2874	2873	2872	2871.5	2871	2870	2869	2868
.77	2867	2866	2865	2864	2863	2862.5	2862	2861	2860	2859
.78	2858	2857	2856	2855	2854	2853	2852	2851	2850	2849
.79	2848	2847	2846	2845	2844	2843.5	2843	2842	2841	2840
.80	2839	2838	2837	2836	2835	2834.5	2834	2833	2832	2831
.81	2830	2829	2828	2827	2826	2825	2824	2823	2822	2821
.82	2820	2819	2818	2817	2816	2815.5	2815	2814	2813	2812
.83	2811	2810	2809	2808	2807	2806	2805	2804	2803	2802
.84	2801	2800	2799	2798	2797	2796.5	2796	2795	2794	2793
.85	2792	2791	2790	2789	2788	2787.5	2787	2786	2785	2784
.86	2783	2782	2781	2780	2779	2778	2777	2776	2775	2774
.87	2773	2772	2771	2770	2769	2768.5	2768	2767	2766	2765
.88	2764	2763	2762	2761	2760	2759.5	2759	2758	2757	2756
.89	2755	2754	2753	2752	2751	2750.5	2750	2749	2748	2747
.90	2746	2745	2744	2743	2742	2741	2740	2739	2738	2737
.91	2736	2735	2734	2733	2732	2731	2730	2729	2728	2727
.92	2726	2725	2724	2723	2722	2721.5	2721	2720	2719	2718
.93	2717	2716	2715	2714	2713	2712.5	2712	2711	2710	2709
.94	2708	2707	2706	2705	2704	2703.5	2703	2702	2701	2700
.95	2699	2698	2697	2696	2695	2694	2693	2692	2691	2690
.96	2689	2688	2687	2686	2685	2684.5	2684	2683	2682	2681
.97	2680	2679	2678	2677	2676	2675.5	2675	2674	2673	2672
.98	2671	2670	2669	2668	2667	2666.5	2666	2665	2664	2663
27.99	2662	2661	2660	2659	2658	2657	2656	2655	2654	2653
In.dec.	0	1	2	3	4	5	6	7	8	9

In.dec.	In. dec. 28.000					In. dec. 28.499				
	0	1	2	3	4	5	6	7	8	9
28.00	2652	2651	2650	2649	2648	2647.5	2647	2646	2645	2644
.01	2643	2642	2641	2640	2639	2638	2637	2636	2635	2634
.02	2633	2632	2631	2630	2629	2628.5	2628	2627	2626	2625
.03	2624	2623	2622	2621	2620	2619.5	2619	2618	2617	2616
.04	2615	2614	2613	2612	2611	2610	2609	2608	2607	2606
.05	2605	2604	2603	2602	2601	2600.5	2600	2599	2598	2597
.06	2596	2595	2594	2593	2592	2591.5	2591	2590	2589	2588
.07	2587	2586	2585	2584	2583	2582.5	2582	2581	2580	2579
.08	2578	2577	2576	2575	2574	2573.5	2573	2572	2571	2570
.09	2569	2568	2567	2566	2565	2564	2563	2562	2561	2560
.10	2559	2558	2557	2556	2555	2554.5	2554	2553	2552	2551
.11	2550	2549	2548	2547	2546	2545	2544	2543	2542	2541
.12	2540	2539	2538	2537	2536	2535.5	2535	2534	2533	2532
.13	2531	2530	2529	2528	2527	2526.5	2526	2525	2524	2523
.14	5522	2521	2520	2519	2518	2517.5	2517	2516	2515	2514
.15	2513	2512	2511	2510	2509	2508.5	2508	2507	2506	2505
.16	2504	2503	2502	2501	2500	2499	2498	2497	2496	2495
.17	2494	2493	2492	2491	2490	2489.5	2489	2488	2487	2486
.18	2485	2484	2483	2482	2481	2480.5	2480	2479	2478	2477
.19	2476	2475	2474	2473	2472	2471.5	2471	2470	2469	2468
.20	2467	2466	2465	2464	2463	2462.5	2462	2461	2460	2459
.21	2458	2457	2456	2455	2454	2453	2452	2451	2450	2449
.22	2448	2447	2446	2445	2444	2443.5	2443	2442	2441	2440
.23	2439	2438	2437	2436	2435	2434.5	2434	2433	2432	2431
.24	2430	2429	2428	2427	2426	2425	2424	2423	2422	2421
28.25	2420	2419	2418	2417	2416	2415.5	2415	2414	2413	2412
.26	2411	2410	2409	2408	2407	2406.5	2406	2405	2404	2403
.27	2402	2401	2400	2399	2398	2397.5	2397	2396	2395	2394
.28	2393	2392	2391	2390	2389	2388.5	2388	2387	2386	2385
.29	2384	2383	2382	2381	2380	2379	2378	2377	2376	2375
.30	2374	2373	2372	2371	2370	2369.5	2369	2368	2367	2366
.31	2365	2364	2363	2362	2361	2360.5	2360	2359	2358	2357
.32	2356	2355	2354	2353	2352	2351.5	2351	2350	2349	2348
.33	2347	2346	2345	2344	2343	2342.5	2342	2341	2340	2339
34	2338	2337	2336	2335	2334	2333.5	2333	2332	2331	2330
.35	2329	2328	2327	2326	2325	2324	2323	2322	2321	2320
.36	2319	2318	2317	2316	2315	2314.5	2314	2313	2312	2311
.37	2310	2309	2308	2307	2306	2305.5	2305	2304	2303	2302
.38	2301	2300	2299	2298	2297	2296	2295	2294	2293	2292
39	2291	2290	2289	2288	2287	2286.5	2286	2285	2284	2283
.40	2282	2281	2280	2279	2278	2277.5	2277	2276	2275	2274
.41	2273	2272	2271	2270	2269	2268.5	2268	2267	2266	2265
.42	2264	2263	2262	2261	2260	2259.5	2259	2258	2257	2256
.43	2255	2254	2253	2252	2251	2250.5	2250	2249	2248	2247
.44	2246	2245	2244	2243	2242	2241.5	2241	2240	2239	2238
.45	2237	2236	2235	2234	2233	2232.5	2232	2231	2230	2229
.46	2228	2227	2226	2225	2224	2223	2222	2221	2220	2219
.47	2218	2217	2216	2215	2214	22 3.5	2213	2212	2211	2210
.48	2209	2208	2207	2206	2205	2204.5	2204	2203	2202	2201
28.49	2200	2199	2198	2197	2196	2195.5	2195	2194	2193	2192
In.dec.	0	1	2	3	4	5	6	7	8	9

	In. dec. 28.500					In. dec. 28.999			
In.dec.	0	1	2	3	4	5	6	7	8	9
28.50	2191	2190	2189	2188	2187	2186.5	2186	2185	2184	2183
.51	2182	2181	2180	2179	2178	2177.5	2177	2176	2175	2174
.52	2173	2172	2171	2170	2169	2168.5	2168	2167	2166	2165
.53	2164	2163	2162	2161	2160	2159.5	2159	2158	2157	2156
.54	2155	2154	2153	2152	2151	2150	2149	2148	2147	2146
.55	2145	2144	2143	2142	2141	2140.5	2140	2139	2138	2137
.56	2136	2135	2134	2133	2132	2131.5	2131	2130	2129	2128
.57	2127	2126	2125	2124	2123	2122.5	2122	2121	2120	2119
.58	2118	2117	2116	2115	2114	2113.5	2113	2112	2111	2110
.59	2109	2108	2107	2106	2105	2104	2103	2102	2101	2100
.60	2099	2098	2097	2096	2095	2094.5	2094	2093	2092	2091
.61	2090	2089	2088	2087	2086	2085.5	2085	2084	2083	2082
.62	2081	2080	2079	2078	2077	2076.5	2076	2075	2074	2073
.63	2072	2071	2070	2069	2068	2067.5	2067	2066	2065	2064
.64	2063	2062	2061	2060	2059	2058.5	2058	2057	2056	2055
.65	2054	2053	2052	2051	2050	2049.5	2049	2048	2047	2046
.66	2045	2044	2043	2042	2041	2040.5	2040	2039	2038	2037
.67	2036	2035	2034	2033	2032	2031.5	2031	2030	2029	2028
.68	2027	2026	2025	2024	2023	2022.5	2022	2021	2020	2019
.69	2018	2017	2016	2015	2014	2013.5	2013	2012	2011	2010
.70	2009	2008	2007	2006	2005	2004.5	2004	2003	2002	2001
.71	2000	1999	1998	1997	1996	1995.5	1995	1994	1993	1992
.72	1991	1990	1989	1988	1987	1986	1985	1984	1983	1982
.73	1981	1980	1979	1978	1977	1976.5	1976	1975	1974	1973
.74	1972	1971	1970	1969	1968	1967.5	1967	1966	1965	1964
28.75	1963	1962	1961	1960	1959	1958.5	1958	1957	1956	1955
.76	1954	1953	1952	1951	1950	1949.5	1949	1948	1947	1946
.77	1945	1944	1943	1942	1941	1940.5	1940	1939	1938	1937
.78	1936	1935	1934	1933	1932	1931.5	1931	1930	1929	1928
.79	1927	1926	1925	1924	1923	1922.5	1922	1921	1920	1919
.80	1918	1917	1916	1915	1914	1913.5	1913	1912	1911	1910
.81	1909	1908	1907	1906	1905	1904.5	1904	1903	1902	1901
.82	1900	1899	1898	1897	1896	1895.5	1895	1894	1893	1892
.83	1891	1890	1889	1888	1887	1886.5	1886	1885	1884	1883
.84	1882	1881	1880	1879	1878	1877.5	1877	1876	1875	1874
.85	1873	1872	1871	1870	1869	1868.5	1868	1867	1866	1865
.86	1864	1863	1862	1861	1860	1859.5	1859	1858	1857	1856
.87	1855	1854	1853	1852	1851	1850.5	1850	1849	1848	1847
.88	1846	1845	1844	1843	1842	1841.5	1841	1840	1839	1838
.89	1837	1836	1835	1834	1833	1832.5	1832	1831	1830	1829
.90	1828	1827	1826	1825	1824	1823.5	1823	1822	1821	1820
.91	1819	1818	1817	1816	1815	1814.5	1814	1813	1812	1811
.92	1810	1809	1808	1807	1806	1805.5	1805	1804	1803	1802
.93	1801	1800	1799	1798	1797	1796.5	1796	1795	1794	1793
.94	1792	1791	1790	1789	1788	1787.5	1787	1786	1785	1784
.95	1783	1782	1781	1780	1779	1778.5	1778	1777	1776	1775
.96	1774	1773	1772	1771	1770	1769.5	1769	1768	1767	1766
.97	1765	1764	1763	1762	1761	1760.5	1760	1759	1758	1757
.98	1756	1755	1754	1753	1752	1751.5	1751	1750	1749	1748
28.99	1747	1746	1745	1744	1743	1742.5	1742	1741	1740	1739
In.dec.	0	1	2	3	4	5	6	7	8	9

(n.dec.)	In dec. 29.000					In dec. 29.499				
	0	1	2	3	4	5	6	7	8	9
29.00	1738	1737	1736	1735	1734	1733.5	1733	1732	1731	1730
.01	1729	1728	1727	1726	1725	1724.5	1724	1723	1722	1721
.02	1720	1719	1718	1717	1716	1715.5	1715	1714	1713	1712
.03	1711	1710	1709	1708	1707	1706.5	1706	1705	1704	1703
.04	1702	1701	1700	1699	1698	1697.5	1697	1696	1695	1694
.05	1693	1692	1691	1690	1689	1688.5	1688	1687	1686	1685
.06	1684	1683	1682	1681	1680	1679.5	1679	1678	1677	1676
.07	1675	1674	1673	1672	1671	1670.5	1670	1669	1668	1667
.08	1666	1665	1664	1663	1662	1661.5	1661	1660	1659	1658
.09	1657	1656	1655	1654	1653	1652.5	1652	1651	1650	1649
.10	1648	1647	1646	1645	1644	1643.5	1643	1642	1641	1640
.11	1639	1638	1637	1636	1635	1634.5	1634	1633	1632	1631
.12	1630	1629	1628	1627	1626	1625.5	1625	1624	1623	1622
.13	1621	1620	1619	1618	1617	1616.5	1616	1615	1614	1613
.14	1612	1611	1610	1609	1608	1607.5	1607	1606	1605	1604
.15	1603	1602	1601	1600	1599	1598.5	1598	1597	1596	1595
.16	1594	1593	1592	1591	1590	1589.5	1589	1588	1587	1586
.17	1585	1584	1583	1582	1581	1580.5	1580	1579	1578	1577
.18	1576	1575	1574	1573.5	1573	1572.5	1572	1571	1570	1569
.19	1568	1567	1566	1565	1564	1563.5	1563	1562	1561	1560
.20	1559	1558	1557	1556	1555	1554.5	1554	1553	1552	1551
.21	1550	1549	1548	1547	1546	1545.5	1545	1544	1543	1542
.22	1541	1540	1539	1538	1537	1536.5	1536	1535	1534	1533
.23	1532	1531	1530	1529	1528	1527.5	1527	1526	1525	1524
.24	1523	1522	1521	1520	1519	1518.5	1518	1517	1516	1515
29.25	1514	1513	1512	1511	1510	1509.5	1509	1508	1507	1506
.26	1505	1504	1503	1502	1501	1500.5	1500	1499	1498	1497
.27	1496	1495	1494	1493	1492	1491.5	1491	1490	1489	1488
.28	1487	1486	1485	1484	1483	1482.5	1482	1481	1480	1479
.29	1478	1477	1476	1475	1474	1473.5	1473	1472	1471	1470
.30	1469	1468	1467	1466	1465	1464.5	1464	1463	1462	1461
.31	1460	1459	1458	1457.5	1457	1456.5	1456	1455	1454	1453
.32	1452	1451	1450	1449	1148	1447.5	1447	1446	1445	1444
.33	1443	1442	1441	1440	1439	1438.5	1438	1437	1436	1435
.34	1434	1433	1432	1431	1430	1429.5	1429	1428	1427	1426
.35	1425	1424	1423	1422	1421	1420.5	1420	1419	1418	1417
.36	1416	1415	1414	1413.5	1413	1412.5	1412	1411	1410	1409
.37	1408	1407	1406	1405	1404	1403.5	1403	1402	1401	1400
.38	1399	1398	1397	1396	1395	1394.5	1394	1393	1392	1391
.39	1390	1389	1388	1387	1386	1385.5	1385	1384	1383	1382
.40	1381	1380	1379	1378	1377	1376.5	1376	1375	1374	1373
.41	1372	1371	1370	1369	1368	1367.5	1367	1366	1365	1364
.42	1363	1362	1361	1360	1359	1358.5	1358	1357	1356	1355
.43	1354	1353	1352	1351	1350	1349.5	1349	1348	1347	1346
.44	1345	1344	1343	1342	1341	1340.5	1340	1339	1358	1337
.45	1336	1335	1334	1333.5	1333	1332.5	1332	1331	1330	1329
.46	1328	1327	1326	1325	1324	1323.5	1323	1322	1321	1320
.47	1319	1318	1317	1316	1315	1314.5	1314	1313	1312	1311
.48	1310	1309	1308	1307	1306	1305.5	1305	1304	1303	1302
29.49	1301	1300	1299	1298	1297	1296.5	1296	1295	1294	1293
(n.dec.)	0	1	2	3	4	5	6	7	8	9

	In. dec. 29.500				In. dec. 29.999				
in.dec.	0	1	2	3	4	5	6	7	8	9
29.50	1292	1291	1290	1289	1288	1287.5	1287	1286	1285	1284
.51	1283	1282	1281	1280	1279	1278.5	1278	1277	1276	1275
.52	1274	1273	1272	1271.5	1271	1270.5	1270	1269	1268	1267
.53	1266	1265	1264	1263	1262	1261.5	1261	1260	1259	1258
.54	1257	1256	1255	1254	1253	1252.5	1252	1251	1250	1249
.55	1248	1247	1246	1245.5	1245	1244.5	1244	1243	1242	1241
.56	1240	1239	1238	1237	1236	1235.5	1235	1234	1233	1232
.57	1231	1230	1229	1228	1227	1226.5	1226	1225	1224	1223
.58	1222	1221	1220	1219	1218	1217.5	1217	1216	1215	1214
.59	1213	1212	1211	1210	1209	1208.5	1208	1207	1206	1205
.60	1204	1203	1202	1201	1200	1199.5	1199	1198	1197	1196
.61	1195	1194	1193	1192	1191	1190.5	1190	1189	1188	1187
.62	1186	1185	1184	1183.5	1183	1182.5	1182	1181	1180	1179
.63	1178	1177	1176	1175	1174	1173.5	1173	1172	1171	1170
.64	1169	1168	1167	1166	1165	1164.5	1164	1163	1162	1161
.65	1160	1159	1158	1157	1156	1155.5	1155	1154	1153	1152
.66	1151	1150	1149	1148	1147	1146.5	1146	1145	1144	1143
.67	1142	1141	1140	1139.5	1139	1138.5	1138	1137	1136	1135
.68	1134	1133	1132	1131	1130	1129.5	1129	1128	1127	1126
.69	1125	1124	1123	1122	1121	1120.5	1120	1119	1118	1117
.70	1116	1115	1114	1113.5	1113	1112.5	1112	1111	1110	1109
.71	1108	1107	1106	1105	1104	1103.5	1103	1102	1101	1100
.72	1099	1098	1097	1096	1095	1094.5	1094	1093	1092	1091
.73	1090	1089	1088	1087	1086	1085.5	1085	1084	1083	1082
.74	1081	1080	1079	1078	1077	1076.5	1076	1075	1074	1073
29.75	1072	1071	1070	1069.5	1069	1068.5	1068	1067	1066	1065
.76	1064	1063	1062	1061	1060	1059.5	1059	1058	1057	1056
.77	1055	1054	1053	1052	1051	1050.5	1050	1049	1048	1047
78	1046	1045	1044	1043	1042	1041.5	1041	1040	1039	1038
79	1037	1036	1035	1034	1033	1032.5	1032	1031	1030	1029
.80	1028	1027	1026	1025.5	1025	1024.5	1024	1023	1022	1021
.81	1020	1019	1018	1017	1016	1015.5	1015	1014	1013	1012
.82	1011	1010	1009	1008.5	1008	1007.5	1007	1006	1005	1004
.83	1003	1002	1001	1000	999	998.5	998	997	996	995
.84	994	993	992	991	990	989.5	989	988	987	986
.85	985	984	983	982	981	980.5	980	979	978	977
.86	976	975	974	973	972	971.5	971	970	969	968
.87	967	966	965	964.5	964	963.5	963	962	961	960
.88	959	958	957	956	955	954.5	954	953	952	951
.89	950	949	948	947	946	945.5	945	944	943	942
.90	941	940	939	938	937	936.5	936	935	934	933
.91	932	931	930	929.5	929	928.5	928	927	926	925
.92	924	923	922	921	920	919.5	919	918	917	916
.93	915	914	913	912.5	912	911.5	911	910	909	908
.94	907	906	905	904	903	902.5	902	901	900	899
.95	898	897	896	895	894	893.5	893	892	891	890
.96	889	888	887	886	885	884.5	884	883	882	881
.97	880	879	878	877.5	877	876.5	876	875	874	873
.98	872	871	870	869	868	867.5	867	866	865	864
29.99	863	862	861	860	859	858.5	858	857	856	855
in.dec.	0	1	2	3	4	5	6	7	8	9

	In. dec. 30 000					In. dec. 30 499				
In.dec.	0	1	2	3	4	5	6	7	8	9
30.00	854	853	852	851	850	849.5	849	848	847	846
.01	845	844	843	842.5	842	841.5	841	840	839	838
.02	837	836	835	834	833	832.5	832	831	830	829
.03	828	827	826	825.5	825	824.5	824	823	822	821
.04	820	819	818	817	816	815.5	815	814	813	812
.05	811	810	809	808	807	806.5	806	805	804	803
.06	802	801	800	799.5	799	798.5	798	797	796	795
.07	794	793	792	791	790	789.5	789	788	787	786
.08	785	784	783	782	781	780.5	780	779	778	777
.09	776	775	774	773	772	771.5	771	770	769	768
.10	767	766	765	764.5	764	763.5	763	762	761	760
.11	759	758	757	756.5	756	755.5	755	754	753	752
.12	751	750	749	748	747	746.5	746	745	744	743
.13	742	741	740	739	738	737.5	737	736	735	734
.14	733	732	731	730	729	728.5	728	727	726	725
.15	724	723	722	721.5	721	720.5	720	719	718	717
.16	716	715	714	713	712	711.5	711	710	709	708
.17	707	706	705	704	703	702.5	702	701	700	699
.18	698	697	696	695.5	695	694.5	694	693	692	691
.19	690	689	688	687	686	685.5	685	684	683	682
.20	681	680	679	678.5	678	677.5	677	676	675	674
.21	673	672	671	670	669	668.5	668	667	666	665
.22	664	663	662	661	660	659.5	659	658	657	656
.23	655	654	653	652.5	652	651.5	651	650	649	648
.24	647	646	645	644	643	642.5	642	641	640	639
30.25	638	637	636	635	634	633.5	633	632	631	630
.26	629	628	627	626.5	626	625.5	625	624	623	622
.27	621	620	619	618	617	616.5	616	615	614	613
.28	612	611	610	609.5	609	608.5	608	607	606	605
.29	604	603	602	601.5	600	599.5	599	598	597	596
.30	595	594	593	592	591	590.5	590	589	588	587
.31	586	585	584	583.5	583	582.5	582	581	580	579
.32	578	577	576	575	574	573.5	573	572	571	570
.33	569	568	567	566	565	564.5	564	563	562	561
.34	560	559	558	557.5	557	556.5	556	555	554	553
.35	552	551	550	549.5	549	548.5	548	547	546	545
.36	544	543	542	541	540	539.5	539	538	537	536
.37	535	534	533	532	531	530.5	530	529	528	527
.38	526	525	524	523.5	523	522.5	522	521	520	519
.39	518	517	516	515	514	513.5	513	512	511	510
.40	509	508	507	506	505	504.5	504	503	502	501
.41	500	499	498	497.5	497	496.5	496	495	494	493
.42	492	491	490	489.5	489	488.5	488	487	486	485
.43	484	483	482	481	480	479.5	479	478	477	476
.44	475	474	473	472	471	470.5	470	469	468	467
.45	466	465	464	463.5	463	462.5	462	461	460	459
.46	458	457	456	455	454	453.5	453	452	451	450
.47	449	448	447	446.5	446	445.5	445	444	443	442
.48	441	440	439	438	437	436.5	436	435	434	433
30.49	432	431	430	429.5	429	428.5	428	427	426	425
In.dec.	0	1	2	3	4	5	6	7	8	9

	In. dec. 30.500					In. dec. 30 999				
In.dec.	0	1	2	3	4	5	6	7	8	9
30.50	424	423	422	421	420	419	418	417	416	415.5
.51	415	414	413	412	411	410	409.5	409	408.5	408
.52	407	406	405	404	403	402	401	400.5	400	399
.53	398	397	396	395	394	393	392.5	392	391	390
.54	389	388	387	386.5	386	385.5	385	384	383	382
.55	381	380	379.5	379	378.5	378	377	376	375	374
.56	373	372	371.5	371	370	369	368	367	366	365
.57	364	363	362.5	362	361	360	359	358	357	356
.58	355	354	353.5	353	352.5	352	351	350	349	348
.59	347	346	345	344	343.5	343	342	341	340	339
.60	338	337.5	337	336	335.5	335	334	333	332	331
.61	330	329	328	327	326.5	326	325	324	323	322
.62	321	320	319	318.5	318	317.5	317	316	315	314
.63	313	312	311	310	309.5	309	308	307	306	305
.64	304	303	302	301.5	301	300	299	298	297	296
.65	295	294	293	292.5	292	291.5	291	290	289	288
.66	287	286	285	284.5	284	283.5	283	282	281	280
.67	279	278	277	276	275	274.5	274	273	272	271
.68	270	269	268	267.5	267	266.5	266	265	264	263
.69	262	261	260	259	258	257.5	257	256	255	254
70	253	252	251	250.5	250	249.5	249	248	247	246
71	245	244	243	242	241	240.5	240	239	238	237
72	236	235	234	233.5	233	232.5	232	231	230	229
73	228	227	226	225.5	225	224.5	224	223	222	221
.74	220	219	218	217	216	215.5	215	214	213	212
30.75	211	210	209	208	207	206.5	206	205	204	203
.76	202	201	200	199.5	199	198.5	198	197	196	195
.77	194	193	192	191	190	189.5	189	188	187	186
.78	185	184	183	182.5	182	181.5	181	180	179	178
.79	177	176	175	174.5	174	173.5	173	172	171	170
.80	169	168	167	166	165	164.5	164	163	162	161
.81	160	159	158	157.5	157	156.5	156	155	154	153
.82	152	151	150	149	148	147.5	147	146	145	144
.83	143	142	141	140.5	140	139.5	139	138	137	136
.84	135	134	133	132	131	130.5	130	129	128	127
.85	126	125	124	123.5	123	122.5	122	121	120	119
.86	118	117	116	115	114	113.5	113	112	111	110
.87	109	108	107	106	105	104.5	104	103	102	101
.88	100	99	98	97.5	97	96.5	96	95	94	93
.89	92	91	90	89.5	89	88.5	88	87	86	85
.90	84	83	82	81.5	81	80.5	80	79	78	77
.91	76	75	74	73	72	71.5	71	70	69	68
.92	67	66	65	64.5	64	63.5	63	62	61	60
.93	59	58	57	56	55	54.5	54	53	52	51
.94	50	49	48	47.5	47	46.5	46	45	44	43
.95	42	41	40	39.5	39	38.5	38	37	36	35
.96	34	33	32	31	30	29.5	29	28	27	26
.97	25	24	23	22.5	22	21.5	21	20	19	18
.98	17	16	15	14	13	12.5	12	11	10	9
30.99	8	7	6	5.5	5	4.5	4	3	2	1
In.dec.	0	1	2	3	4	5	6	7	8	9

TABLE III.

For the detached Thermometer.

App.height.	0	20	40	60	80	App.height.	0	20	40	60	80
0	0	.08	.1	.2	.3	5000	21.4	21.4	21.5	21.6	21.7
100	.4	.5	.6	7	.8	100	21.8	21.9	22.0	22.0	22.1
200	.8	.9	1.0	1.1	1.2	200	22.2	22.3	22.4	22.5	22.6
300	1.2	1.3	1.4	1.5	1.6	300	22.6	22.7	22.8	22.9	23.0
400	1.7	1.8	1.8	1.9	2.0	400	23.1	23.2	23.2	23.3	23.4
500	2.1	2.2	2.3	2.4	2.5	500	23.5	23.6	23.7	23.8	23.8
600	2.5	2.6	2.7	2.8	2.9	600	23.9	24.0	24.1	24.2	24.3
700	3.0	3.0	3.1	3.2	3.3	700	24.4	24.4	24.5	24.6	24.7
800	3.4	3.5	3.6	3.7	3.8	800	24.8	24.9	25.0	25.0	25.1
900	3.8	3.9	4.0	4.1	4.2	900	25.2	25.3	25.4	25.5	25.5
1000	4.3	4.3	4,4	4.5	4.6	6000	25.6	25.7	25.8	25.9	26.0
100	4.7	4.8	4.9	5.0	5.0	100	26.0	26.1	26.2	26.3	26.4
200	5.1	5.2	5.3	5.4	5.5	200	26.5	26.6	26.7	26.7	26.8
300	5.5	5.6	5.7	5.8	5.9	300	26.9	27.0	27.0	27.1	27.2
400	6.0	6.0	6.1	6.2	6.3	400	27.3	27.4	27.5	27.6	27.7
500	6.4	6.5	6.6	6.7	6.7	500	27.8	27.9	28.0	28.0	28.1
600	6.8	6.9	7.0	7.0	7.1	600	28.2	28.3	28.4	28.5	28.5
700	7.2	7.3	7.4	7.5	7.6	700	28.6	28.7	28.8	28.9	29.0
800	7.7	7.8	7.9	8.0	8.0	800	29.0	29.1	29.2	29.3	29.4
900	8.1	8.2	8.3	8.3	8.4	900	29.5	29.6	29.6	29.7	29.8
2000	8.5	8.6	8.7	8.8	8.9	7000	29.9	30.0	30.1	30.2	30.3
100	9.0	9.0	9.1	9.2	9.3	100	30.3	30.4	30.5	30.6	30.7
200	9.4	9.5	9.6	9.6	9.7	200	30.8	30.8	30.9	31.0	31.1
300	9.8	9.9	10.0	10.0	10.1	300	31.2	31.3	31.4	31.4	31.5
400	10.2	10.3	10.4	10.5	10.6	400	31.6	31.7	31.8	31.9	32.0
500	10.7	10.8	10.8	10.9	11.0	500	32.0	32.1	32.2	32.3	32.4
600	11.1	11.2	11.2	11.3	11.4	600	32.5	32.6	32.6	32.7	32.8
700	11.5	11.6	11.7	11.8	11.9	700	32.9	33.0	33.1	33.2	33.2
800	12.0	12.0	12.1	12.2	12.3	800	33.3	33.4	33.5	33.6	33.7
900	12.4	12.5	12.6	12.6	12.7	900	33.8	33.8	33.9	34.0	34.1
3000	12.8	12.9	13.0	13.0	13.1	8000	34.2	34.3	34.4	34.5	34.5
100	13.2	13.3	13.4	13.5	13.6	100	34.6	34.7	34.8	34.9	34.9
200	13.7	13.8	13.8	13.9	14.0	200	35.0	35.1	35.2	35.3	35.4
300	14.1	14.2	14.3	14.4	14.4	300	35.5	35.5	35.6	35.7	35.8
400	14.5	14.6	14.7	14.8	14.9	400	35.9	36.0	36.0	36.1	36.2
500	15.0	15.0	15.1	15.2	15.3	500	36.3	36.4	36.5	36.6	36.7
600	15.4	15.5	15.6	15.6	15.7	600	36.7	36.8	36.9	37.0	37.1
700	15.8	15.9	16.0	16.0	16.1	700	37.2	37.3	37.3	37.4	37.5
800	16.2	16.3	16.4	16.5	16.6	800	37.6	37.7	37.8	37.9	37.9
900	16.7	16.7	16.8	16.9	17.0	900	38.0	38.1	38.2	38.3	38.4
4000	17.1	17.2	17.3	17.3	17.4	9000	38.5	38.5	38.6	38.7	38.8
100	17.5	17.6	17.7	17.8	17.9	100	38.9	39.0	39.1	39.1	39.2
200	18.0	18.0	18.1	18.2	18.3	200	39.3	39.4	39.5	39.6	39.7
300	18.4	18.5	18.5	18.6	18.7	300	39.7	39.8	39.9	40.0	40.1
400	18.8	18.9	19.0	19.0	19.1	400	40.2	40.3	40.3	40.4	40.5
500	19.2	19.3	19.4	19.5	19.6	500	40.6	40.7	40.8	40.8	40.9
600	19.6	19.7	19.8	19.9	20.0	600	41.0	41.1	41.2	41.3	41.4
700	20.1	20.2	20.3	20.3	20.4	700	41.4	41.5	41.6	41.7	41.8
800	20.5	20.6	20.7	20.8	20.9	800	41.9	42.0	42.0	42.1	42.2
900	21.0	21.0	21.1	21.2	21.3	900	42.3	42.4	42.5	42.6	42.6
App.height.	0	20	40	60	80	App.height.	0	20	40	60	80

F

TABLE III. *Continued.*

App.height.	0	20	40	60	80	App.height.	0	20	40	60	80
10000	42.7	42.8	42.9	43.0	43.1	15000	64.1	64.2	64.3	64.3	64.4
100	43.2	43.2	43.3	43.4	43.5	100	64.5	64.6	64.7	64.8	64.9
200	43.6	43.7	43.8	43.8	43.9	200	64.9	65.0	65.1	65.2	65.3
300	44.0	44.1	44.2	44.3	44.3	300	65.4	65.5	65.5	65.6	65.7
400	44.4	44.5	44.6	44.7	44.8	400	65.8	65.9	66.0	66.1	66.1
500	44.9	45.0	45.0	45.1	45.2	500	66.2	66.3	66.4	66.5	66.6
600	45.3	45.4	45.5	45.5	45.6	600	66.7	66.7	66.8	66.9	67.0
700	45.7	45.8	45.9	46.0	46.1	700	67.1	67.2	67.3	67.3	67.4
800	46.1	46.2	46.3	46.4	46.5	800	67.5	67.6	67.7	67.8	67.9
900	46.6	46.7	46.7	46.8	46.9	905	67.9	68.0	68.1	68.2	68.3
11000	47.0	47.1	47.2	47.2	47.3	16000	68.4	68.5	68.5	68.6	68.7
100	47.4	47.5	47.6	47.7	47.8	100	68.8	68.9	69.0	69.1	69.2
200	47.9	47.9	48.0	48.1	48.2	200	69.2	69.3	69.4	69.5	69.6
300	48.3	48.4	48.5	48.5	48.6	300	69.6	69.7	69.8	69.9	70.0
400	48.7	48.8	48.9	49.0	49.1	400	70.1	70.2	70.2	70.3	70.4
500	49.2	49.2	49.3	49.4	49.5	500	70.5	70.6	70.7	70.8	70.8
600	49.6	49.6	49.7	49.8	49.9	600	70.9	71.0	71.1	71.2	71.3
700	50.0	50.1	50.2	50.3	50.3	700	71.4	71.4	71.5	71.6	71.7
800	50.4	50.5	50.6	50.7	50.8	800	71.8	71.9	72.0	72.0	72.1
900	50.8	50.9	51.0	51.1	51.2	900	72.2	72.3	72.4	72.5	72.6
12000	51.3	51.4	51.4	51.5	51.6	17000	72.6	72.7	72.8	72.9	73.0
100	51.7	51.8	51.9	52.0	52.0	100	73.1	73.2	73.2	73.3	73.4
200	52.1	52.2	52.3	52.4	52.5	200	73.5	73.6	73.7	73.8	73.8
300	52.6	52.6	52.7	52.8	52.9	300	73.9	74.0	74.1	74.2	74.3
400	53.0	53.1	53.1	53.2	53.3	400	74.3	74.4	74.5	74.6	74.7
500	53.4	53.5	53.6	53.6	53.7	500	74.8	74.9	74.9	75.0	75.1
600	53.8	53.9	54.0	54.1	54.2	600	75.2	75.3	75.4	75.5	75.5
700	54.3	54.3	54.4	54.5	54.6	700	75.6	75.7	75.8	75.9	76.0
800	54.7	54.8	54.9	55.0	55.0	800	76.1	76.1	76.2	76.3	76.4
900	55.1	55.2	55.3	55.4	55.5	900	76.5	76.6	76.7	76.7	76.8
13000	55.5	55.6	55.7	55.8	55.9	18000	76.9	77.0	77.1	77.2	77.3
100	56.0	56.1	56.2	56.3	56.4	100	77.3	77.4	77.5	77.6	77.7
200	56.5	56.6	56.7	56.7	56.8	200	77.8	77.9	77.9	78.0	78.1
300	56.9	56.9	57.0	57.1	57.2	300	78.2	78.3	78.4	78.5	78.5
400	57.3	57.3	57.4	57.5	57.6	400	78.6	78.7	78.8	78.9	79.0
500	57.7	57.8	57.9	57.9	58.0	500	79.1	79.2	79.2	79.3	79.4
600	58.1	58.2	58.3	58.4	58.5	600	79.5	79.6	79.6	79.7	79.8
700	58.5	58.6	58.7	58.8	58.9	700	79.9	80.0	80.1	80.2	80.2
800	59.0	59.1	59.2	59.2	59.3	800	80.3	80.4	80.5	80.6	80.7
900	59.4	59.5	59.5	59.6	59.7	900	80.8	80.8	80.9	81.0	81.1
14000	59.8	59.9	60.0	60.1	60.2	19000	81.2	81.3	81.4	81.4	81.5
100	60.2	60.3	60.4	60.5	60.6	100	81.6	81.7	81.8	81.9	82.0
200	60.7	60.8	60.8	60.9	61.0	200	82.0	82.1	82.2	82.3	82.4
300	61.1	61.2	61.3	61.4	61.4	300	82.5	82.6	82.6	82.7	82.8
400	61.5	61.6	61.7	61.8	61.9	400	82.9	83.0	83.1	83.2	83.2
500	62.0	62.1	62.2	62.3	62.4	500	83.3	83.4	83.5	83.6	83.7
600	62.5	62.6	62.6	62.7	62.8	600	83.8	83.8	83.9	84.0	84.1
700	62.9	63.0	63.0	63.1	63.2	700	84.2	84.3	84.3	84.4	84.5
800	63.2	63.3	63.4	63.5	63.6	800	84.6	84.7	84.8	84.9	84.9
900	63.7	63.8	63.8	63.9	64.0	900	85.0	85.1	85.2	85.3	85.4
App.height.	0	20	40	60	80	App.height.	0	20	40	60	80

THE END.

PRINTED BY

RICHARD AND ARTHUR TAYLOR,

SHOE LANE, LONDON.

Printed in the United States
By Bookmasters